Solving Problems in
Technical Writing

Solving Problems in Technical Writing

Edited by

LYNN BEENE and PETER WHITE

University of New Mexico

New York   Oxford
OXFORD UNIVERSITY PRESS
1988

Oxford University Press

Oxford   New York   Toronto
Delhi   Bombay   Calcutta   Madras   Karachi
Petaling Jaya   Singapore   Hong Kong   Tokyo
Nairobi   Dar es Salaam   Cape Town
Melbourne   Auckland

and associated companies in
Berlin   Ibadan

Copyright © 1988 by Oxford University Press, Inc.

Published by Oxford University Press, Inc.
200 Madison Avenue, New York, New York 10016

Oxford is a registered trademark of Oxford University Press

Library of Congress Cataloging-in-Publication Data
Solving problems in technical writing / edited by Lynn Beene and Peter White
p.   cm.   Bibliography: p.   Includes index.
ISBN 0-19-505330-3; ISBN 0-19-505331-1 (pbk.)
1. English language—Rhetoric—Study and teaching.
2. Technical writing—Study and teaching.
I. Beene, Lynn.   II. White, Peter, 1947–.
PE1475.S75 1988
808'.0666'07—dc19   87-30439   CIP

We gratefully acknowledge permission to reprint from the following works:
The Ascent of Man by J. Bronowski. Boston: Little, Brown, 1973. Copyright ©
1973 by J. Bronowski.
Broca's Brain by Carl Sagan. New York: Random House, 1979, p. 37. Copyright
© 1978 by Carl Sagan. All Rights Reserved. Reprinted by permission of the author.
Collected Poems, Volume I: 1909–1939 by William Carlos Williams. New York:
New Directions, 1938. Copyright © 1938 by New Directions Publishing Corporation.
Forms of Talk by Erving Goffman. Philadelphia: University of Pennsylvania Press,
1981, pp. 172, 191, 194–195.
The Mismeasure of Man by Stephen Jay Gould. New York: W. W. Norton,
1981, p. 108. Copyright © 1981 by Stephen Jay Gould.
Mr. Maugham Himself by Somerset Maugham. New York: Doubleday, 1954,
p. 558.

Printing (last digit): 9 8 7 6 5 4 3 2 1

Printed in the United States of America
on acid-free paper

Preface

Several years ago we recognized the need for answers to some of the critical and fundamental problems associated with professional writing. We therefore called upon experts to provide solutions, from both theoretical and practical bases, for dilemmas encountered throughout the composing process—from conception to organization, composition, revision, illustration, and publication. We asked our consultants to address issues and options facing students, technicians, scientific researchers, and practicing professionals. This book, then, has several audiences, all of whom face similar problems in communicating information effectively.

The strengths of this collection are the variety of viewpoints expressed and the combination of theory and practical advice dispensed. There is a growing body of knowledge in this relatively new field that should be made available to all professional writers in concise and readable essays. This collection neither overwhelms students or beginning writers nor condescends to professional technical writers. Thus we hope this book will be used by both professional writers and those who supervise. *Solving Problems in Technical Writing* gives the reader a broader perspective, a sense of the major goals of research and technical documentation.

Throughout the project the editors have benefitted greatly from the advice of the contributors, who were encouraged to take their essays in the directions they saw most appropriate and rewarding for the reader. In addition, this book would not have been possible without the efforts of Rick Eden of the Rand Corporation, Curtis Church, and the staff and editors at Oxford University Press. We would like to acknowledge in particular Mimi Melek and Joan Bossert whose guidance and patience encouraged us at every step. Finally, no words can adequately acknowledge the support and patience David W. Frizzell and Mary Ann White gave us throughout this project.

April 1988
Albuquerque, New Mexico

L. B.
P. W.

Contents

Solving Problems in
Technical Writing

How Can Problem–Solution Structures Help Writers Plan and Write Technical Documents?

MICHAEL P. JORDAN

Selection and organization are two of the major problems writers face when they assemble complex technical documents. Michael P. Jordan, Associate Professor of Technical Communications at Queen's University, argues that writers can improve the organization of their documents and select appropriate information if they adopt a problem solving practice long employed by engineers: surveying the subject, understanding the problem faced, finding or creating a solution, and implementing and evaluating that solution. Following the lead of linguists such as E. O. Winter, Jordan applies this approach to effective functional texts. The name for this rhetorical structure, *Situation–Problem–Solution–Evaluation*, identifies each of the four main elements of that approach. Jordan cites a variety of texts, from engineering reports to formal proposals, as illustrations of this basic structure and its several typical variations.

Jordan shows that, although this four-part structure is often complete, in complex technical documents it can be truncated or may appear in complicated guises. For example, a writer may need to omit a given section because that topic is addressed elsewhere in the report, or a writer may choose to explain detail distant from the report's major topic because of a deficiency in a previous solution. Further complicating the basic four-part structure are demands made by the subject to provide details of the causes, seriousness, or location of a problem, or a variety of possible solutions, or both positive and negative evaluations. Often a writer must issue a progress report or report of a failed project or one halted before successful completion.

Structuring Technical Documents

Because most engineering and scientific activities are essentially problem solving, many of the documents that describe these activities follow definable problem-solving structures. Yet in spite of this, engineers, scientists, and technical writers receive little if any formal instruction in the use of these structures to plan or write technical documents. Recent research and related publications now make it possible to study this subject and then to use the principles in practice. This chapter provides the necessary introduction to this subject, with the aim of assisting writers to improve both their understanding of and ability to use these prevalent forms of technical writing.

This introductory paragraph follows a well-established structure. First, it establishes the background situation of technical documentation being essentially involved with problem solving. Within this situation, a problem is identified in the second sentence: we would expect, in the light of the information in the first sentence, to be told that technical writers are fully aware of the structures of problem-solving documents in English, but we are told that this is not so and we recognize this deficiency as a problem that needs to be remedied. The third sentence introduces information that a solution is now available, and the final sentence explains how the work discussed in this chapter will provide at least a partial solution to the defined needs. The structure of the first paragraph can thus be seen to follow the simple pattern of *Situation–Problem–Solution–Evaluation,* where Evaluation is an indication of the extent to which the proposed Solution is an actual solution to the Problem.

The first paragraph also has interesting signals that guide the reader through the text and illustrate the type of information being conveyed at each point in the text. The use of *Yet* mediates between the situation and the problem, and *in spite of this* indicates the denial of the expected conclusion of the second sentence based on the information given in the first. In the third sentence, *now it is possible* signals that a solution is at hand; the final sentence provides information by which we can judge the possible effectiveness of the suggested *improvement.*

This chapter is a study of the principles and use of information structures, such as *Situation–Problem–Solution–Evaluation,* together with a description of the signals essential to their communication. It addresses the important practical and pedagogical problem of how we account for problem-solving texts within a consistent theoretical framework and how we can use this understanding to plan and write technical documents.

Relevant Research

The Technical Problem Solver

Although the structures and signaling devices discussed here apply equally well to most professional activities, this anthology deals with writing broadly within engineering and science. This chapter provides a detailed account of the use of some established scholarship on both the planning and writing of technical documents.

Engineering is often defined as the application of scientific principles for the betterment of mankind, and many engineering schools prefer the title of "Applied Science" to stress their objectives. Essentially, engineering involves solving problems in its widest sense, including overcoming difficulties, meeting established needs, and improving existing methods. A "design methodology" has been established that attempts to explain the complex series of steps of any problem-solving procedure from conception to fruition. Numerous texts are available on this subject including the work by Edel (1967), which provides some general background, and the design graphics text by Earl (1977), which applies a design methodology for teaching purposes.

Because of the complex nature of the series of steps in engineering problem-solving, analysts have a different number of steps and use different labels for each of the steps. The principles, however, appear common to all: understanding the situation being faced; analyzing the specific problem to be tackled; creating, analyzing, and refining a solution; and further evaluating, improving, and implementing. This, of course, is a very crude model of the complexities of problem-solving activities, but it is a useful starting point.

The problem-solving nature of science is a little different although just as apparent. The scientist's function is to discover principles that describe how and why things in nature work the way they do and then to seek to explain phenomena within these established principles. This is equally true whether we are concerned with natural, earth, medical, or social science, and similar techniques are even being applied (Jordan, 1985a) in linguistic analysis—"the scientific study of language" (Lyons, 1968, p. 1). In all this scientific work, the worker first needs to grasp the general subject matter and then define a specific problem to be solved. The hypotheses based on observations and related scholarship are possible solutions to the scientist's "need-to-know" problem; and their analyses, refinement, and final adoption as scientific "laws" clearly parallel the engineering model.

Technical documentation is no different. Just as an engineering product is designed to meet a specific problem and a scientific explanation is devised to explain a specific set of phenomena, so are all technical documents planned to meet specific needs of the writer and readers. The planning, writing, and producing of any technical document follow the same procedure as the creation and manufacture of any engineering product—a parallel recognized by the well-known writing process.

Problem-Solving Structures in Language

It seems almost too obvious to state that documents that explain how engineers and scientists work must almost inevitably be connected with matters of problem solving. Yet few linguists have analyzed English texts beyond the sentence, and even fewer have analyzed the structures of examples of technical writing at paragraph level and beyond. The result of this lack of specific language-based scholarship in technical writing is that the subject is not yet based (like other subjects) on a sound body of theoretical language knowledge that students and practitioners can understand and apply. The work introduced here should be seen as only a partial answer to this greater need.

Given that the patterns of human thought and action are extremely complicated, we must expect that any account of these thoughts and actions in a technical document will also be exceptionally complicated. Indeed, we might initially feel that analysis of such texts must inevitably fail, as we cannot possibly classify and account for all the possible variations and complications that occur in practical problem-solving activities. But that is not the case. True, these activities as complete sets of thought and action *are* difficult to classify. But the individual subparts of the activities are quite predictable and classifiable. As any text is purely a representational summary of the complex thoughts and actions it describes, an understanding of the structure of texts helps to develop an understanding of the processes involved in the work itself.

Survey of Research and Teaching Material

General Background

The roots of the work described here can be traced to the Communications Research Centre at University College, London, England, which in 1953 was established to study the many forms of educated English use as a research subject of importance in its own right. Under the leadership of M. A. K. Halliday (whose analysis of transitivity and theme (1967) has considerable significance as background for this work), an extensive survey, the "Organisation of Scientific and Technical Information" (OSTI), was conducted into three levels of technical writing (Huddleston et al., 1968; Huddleston, 1971). Of direct relevance for this work is the analysis by E. O. Winter whose work in the OSTI program dealt with systems of cohesion between clauses and sentences in technical writing. This and related work by Winter (1971) served as the basis for three quite distinct but related areas of analysis of technical English:

1. *Clause relations,* which explains the relationships between clauses, sentences, and paragraphs and the linguistic signals that communicate them. This is closely related to work in rhetorical predicates (Grimes, 1975) and relational propositions (Beekman & Callow, 1974; Gross & Sidner, 1986; Mann & Thompson, 1983, 1987). Earlier work by Winter (1977) has been expanded by his analysis of the predictive nature of such words as *reason, results,* and *contrast;* and Jordan (1984b) has recently shown their use in an analysis of clause relations in technical writing. Other work (Jordan, 1985c) includes more detailed analysis of the clause relations of logical connection in technical writing, and further background discussion of clause relations is given by Hoey (1983) and by Jordan (1988a, 1988b).

2. *Lexical cohesion,* which explains continuity in technical writing created by continued "reentry" of the topic of description into new clauses and sentences. Based on earlier work by Christophersen (1939), Hawkins (1978), and Halliday and Hasan (1976) and established systems of coreferential continuity, Jordan has developed the work to describe the genre of technical description (1982a), associated nominals in technical writing (1981a), the thread of continuity in functional writing (1982b), nonthematic reentry (1985b), and complex lexical cohesion (1983). Many of these concepts are related to technical description (Jordan, 1984c).

3. *Rhetorical structures* (also known as "prose structures" or "information structures"), which explains the types of information presented within a text, their sequence and systems of combination, and also the linguistic signals that separate and combine them. This is the area of research that concerns us here.

Rhetorical Structures

The need for research into the structures of technical prose became apparent to Winter in the early 1970s when he was faced with the task of helping Norwegian students to write reports in English. After analyzing hundreds of technical texts, he concluded that many of them followed some sort of pattern explained in general terms by the structure *Situation–Problem–Solution–Evaluation.* Although it appears deceptively simple and perhaps a little naive, this system does provide a meaningful basis for analyzing the structure of a large number of technical texts; Winter used this approach successfully for Norwegian students and later in helping both business and engineering students in England to improve their writing abilities. His pioneering work was partially developed as a manual for teachers and students (Winter, 1976), which has been used extensively in England and other parts of Europe and can be seen as the basis for later work.

Working with Winter, Hoey explained some of the complexities and signaling devices relating to the four types of information (1979) and applied the principles to literary analysis (1981). More importantly, Hoey's textbook analyzes the surface matters of discourse analysis (1983) and contains three chapters discussing and elaborating many of the concepts introduced and explained by Winter. An advance in the theory was established by Jordan (1980), who demonstrated a variety of patterns within the overall framework of *Situation–Problem–Solution–Evaluation* and showed how various types of progress reports and "incomplete" structures can be explained within the same consistent framework of analysis. The principles have also been applied to expository writing (Jordan, 1982c) and to some very short English texts (Jordan, 1981b). In addition, a textbook devoted entirely to the analysis of technical and other texts within the established framework of rhetorical structures is now available (Jordan, 1984d).

The remainder of this section on research background provides a very brief summary of some of the major principles involved. New examples are used here so that this discussion supplements analyses available elsewhere.

The Four-Part Structure in Practice

The best starting point for understanding rhetorical structures is to study texts that clearly present the four major parts of any problem-solving series of events. Here is a brief text of this type:

1. Four-Part Structure

As the structural engineering consultants on Roy Thomson Hall, Carruthers & Wallace were faced with a number of challenges. Acoustics requirements dictated that all interior surfaces must be faceted sufficiently to scatter sound so that all seats in the auditorium receive music of equal clarity and quality. This requirement was satisfied by the use of exposed concrete for all interior surfaces enclosing the auditorium. Precast concrete panels, each set at a slightly different angle, ensure uniform reflection of sound from ceiling surfaces. (*Engineering Dimensions,* Jan./Feb. 1984, p. 10)

These four sentences can be labeled *Situation, Problem, Solution,* and *Evaluation,* respectively. The first sentence provides details of the consultants and their major task together with a harbinger of the problems to come with the words *faced with a number of challenges.* One of these is mentioned in the second sentence in terms of the need dictated by design requirements. This need is epitomized by *This requirement* in the third sentence, and the words *This requirement was satisfied by* both indicate that a solution is about to be given and also that it actually *is* a solution as the need has been *satisfied.* The *Solution* is thus "preevaluated" as being an effective solution even before it is detailed in the text. Details of the *Solution* are provided in the remainder of the third sentence, and the means by which it meets the sound requirement are given in the last sentence. The word *ensure* is a significant evaluative word here coupled with the rest of the sentence. Although this example is part of a much larger text and the thoughts and actions it represents are only a small part of a much larger problem-solving series of activities, we can still understand the pattern in terms of *Situation–Problem–Solution–Evaluation.* We can also see that, as with all such complex procedures, problem–solution thought patterns appear within larger problem–solution patterns: larger texts can contain smaller structures within them. Hoey (1983) discusses this in some detail.

What is significant about the four-part structure is that it provides a satisfyingly complete account (or "story" if you like) of the thoughts and actions of the engineers or scientists involved in the work. A text that does not contain these four parts could leave the readers unsatisfied, wondering whether the work has yet to be completed. There are reasons for not including some parts of the overall structure, however, and these are discussed later.

Some Complexities

In practice much more may be involved in solving a problem than simply creating or discovering a suitable solution to a problem; thus the four-part structure could be misleading in its simplicity. Some of the complexities that occur in practice are represented in the following text:

2. Floating Error

When Avco Engineering were informed by one of their customers that a precision jig that Avco had supplied was inaccurate, the company sent engineers to check. The customer was an aircraft manufacturer, and the jig was laid on the concrete floor of a hangar on a river estuary in Somerset.

The acceptable tolerance was 0.003 inches over 15 ft (0.08 mm over 4.6 m) and, when a check was made the following morning, the jig was found to be well within this limit. After lunch, however, when Avco engineers were endeavouring to demonstrate this fact to engineers from the aircraft company, a discrepancy of 0.02 inches (0.5 mm) was measured! Subsequent checks during the course of the day revealed declining inaccuracies. Surveys carried out the next day revealed a similar pattern.

It was not until the engineers turned their attention to the base on which the jig was laid—the hangar floor—that the mystery was solved. The Avco man demonstrated that, as the tide rose and fell in the estuary, the concrete floor distorted, causing the jig to move. Once the cause was discovered, the solution was simple—the jig was re-sited on a more stable base. (*Chartered Mechanical Engineer,* Nov. 1978, p. 24)

Here again the solution was obvious and successful, but it could not be arrived at until first the engineers realized that there was a problem, the extent of the problem, and then the specific causes of the problem. The *Situation* and *Problem* are included in the first paragraph, the word *inaccurate* being a clear signal of *Problem*. (In fact most prefixes are *Problem* signaling as they negate a desirable attribute, Jordan, 1984d.) We are told that at first Avco engineers did not recognize a problem as the jig was found to be well within the limit. So convinced were they that no problem existed that they tried to demonstrate this to their customer, only to find a *discrepancy* (obviously a problem) outside the acceptable limit. Further work was necessary, and this revealed the cyclic pattern of change in tolerance and the conclusion that the tide was the cause of the problem. (Note that the lack of knowledge of the cause of the discrepancy was itself seen as a problem and that this mystery was solved by the understanding of the cause—another example of a *Problem–Solution* structure within a larger one.) The final sentence tells us about the success of the obvious solution (preevaluated as *the solution was simple*) once the cause of the problem was apparent.

Example 2 demonstrates some of the complexities that can be included within the section of *Problem* within a text. There may need to be details of whether there is a problem, its cause, extent, seriousness, solvability, or location—and there may be several related problems that have to be considered. For very complex projects it may be necessary to decide which of several problems should have effort and money expended in attempting to overcome them, and problem definitions and analyses can become extremely complex. We should now be able to recognize that performance specifications are defined statements of problems in terms of the design requirements of a product and that material-and-construction specifications are statements of the solution to a problem.

Complications with Solutions

Just as we can expect in a larger report to have information giving complex details of the problem, so too can we often expect complex details under the general concept of *Solution*. Here is the introduction to a report illustrating some of these complexities:

3. Vertical Drainage

One of the oldest and most accepted methods of strengthening weak soils is to preload the ground before construction, hastening compression and drainage by using vertical drains. For many years the preeminent vertical drainage technique was the emplacement of sand drains.

A sand drain is a vertical column of sand, usually 200 to 500 mm in diameter, which provides an escape route for groundwater squeezed out by a heavy embankment at the surface. While popular, they pose disadvantages: they require large amounts of water, which may not always be available; the specifications for the sand lie within a narrow range; and installation is labor-intensive. Some soil experts believe conventional sand drains installed for accelerating consolidation may soon be a thing of the past. They are being replaced with a wide variety of prefabricated drains.

Prefab or "wick" drains, as they are commonly called, were first developed in Sweden in 1939, about the same time as sand drains came on the scene. The Kjellman cardboard wick, named for its inventor, Walter Kjellman, was widely used for years as an alternative to sand drains, especially in Japan. The wick consisted of three layers of cardboard—two filters surrounding a rigid interior with grooved channels. Water traveled up the channels in much the same way it would percolate through a sand drain.

Almost everything was right about the Kjellman drain, except its material. Cardboard didn't supply enough strength and durability, and the paper filters offered poor permeability. With the advent of plastics, however, wick drains could be made with more durable cores, and with stronger and more permeable polyester filters. Plastic wicks today can be inserted to depths greater than 100 feet. (*Compressed Air,* Mar. 1983, p. 33)

This example is different from the others given so far in that the *Problem* identification is in terms of a recognized deficiency in an existing solution. First, we are given a general solution to a quite complex problem. The initial problem is the *weak soils,* and their *strengthening* is the obvious solution. The old means of achieving this was the use of sand drains, and details of sand drains are given, followed by their *disadvantages.* Note that whereas *however* was used to mediate between "no problem" and "problem" in Example 2, the subordinator *While* is used here to indicate the "knownness" of the popularity as well as mediating between the good evaluation and the bad evaluation (*Problem*). Details of the problem with sand drains are given after the colon, followed by two evaluations in that second paragraph: one being negative for the sand drains and the other being positive for the prefabricated drain to be discussed in the report. The third paragraph describes the prefab drains and their applications, and the final paragraph provides *Evaluation*. The statement that *almost everything was right* is clearly a positive evaluation, but a problem area (the material) is then identified. The specific problem with material selection is given in the second sentence of the final paragraph with the clear problem negation signal *n't . . . enough*. With *however* mediating between the positive and negative evaluations, the problem of material selection (a problem with a solution) is seen to be overcome by the selection of plastic instead.

The paragraphing of Example 3 exemplifies the *Situation–Problem–Solution–Evaluation* structure, but within each of these parts (particularly the second and fourth paragraphs) we have to recognize other complex structures at work.

The improvement of an existing solution is illustrated below:

4. Sandhu Machine Design

The Rhino XR-1 high-tech robot is a 32-in.-tall machine that is completely open and observable. It uses the same tasks on scale as larger units. However, it is meant to be handled, studied, and even taken apart and reassembled by the user. A tool kit is provided to make the task easy. It is intended primarily as a teaching tool to determine the feasibility of using larger units on production lines. (*Mechanical Engineer,* Aug. 1982, p. 16)

This text could have been written in the form of situational background, followed by the existing solutions and their defects for teaching purposes and by a discussion of this new device and its advantages. However, the writer chose to stress similarities and differences instead. The text starts with details of the *Solution* and then indicates the items of similarity with large systems, *same* being used to indicate this similarity. *However* mediates between the similarities and differences, and the differences are clear in the text that follows. Some important concepts are involved here. First, with any improvement there are bound to be similarities and differences, but it is only the differences that can be seen to overcome the deficiencies in the former method or equipment. Second, similarity and difference involve relations between clauses (part of the study of clause relations), and there is a very rich vocabulary in English to communicate these; Jordan's recent work (1984e) is a detailed discussion of exemplification of the signaling systems used in technical English. Third, the comparison also involves relating information about two or more things within a set (here robots), and this involves the theory of textual cohesion between lexical (or word) choices that are cohyponyms of a superordinate term (Jordan,

1983). Thus, such examples as these are proving extremely useful in current attempts to integrate several branches of research in technical writing.

"Incomplete" Texts

The theory of rhetorical structures can be extended greatly when we realize that the four parts of the overall structure may not occur in the text. There are essentially three reasons for this:

- The work may be stopped prior to a successful solving of a recognized problem. No problem may have been identified; the problem may be deemed not serious enough to warrant remedial action; the problem may be unsolvable with present knowledge; or the project may simply have run out of money or have been beaten by a better solution from a competitor.
- The work may still be in progress at any stage of the process, and the document will then be a progress report of work done to date, perhaps also with predictions for further work. Both possibilities 1 and 2 are described in detail in Jordan (1980) in terms of a multistage algorithm which explains such structures as *Situation–Evaluation, Situation–Problem–Problem Accepted, Situation–Problem–Problem Being Investigated, Situation–Problem–Proposed Solutions Being Tested,* etc., leading eventually to the final structure *Situation–Problem–Solution–Evaluation.*
- For either complete or incomplete activities, writers may not include some parts (usually *Situation* and *Evaluation,* but sometimes also *Problem*) as they feel readers will know the information already or the information will be obvious from the other details given in the text.

As all engineering products are designed to solve a problem, even descriptive material can be seen as information about a *Solution* with perhaps a hint at a problem or need (applications) and maybe also some evaluation in comparison with other products (other solutions). Even products arrived at almost by accident are viewed as solutions—even though the problems they solve have not yet been fully discovered:

5. Decomposable Al-Alloy

A unique Al-alloy (Series 300), developed by TAFA Inc., Bow, N.H., has inherent metal properties such as electrical conductivity, machineability, strength, and durability, but can be rapidly dissolved by water. Series 300 alloy is a byproduct of the company's research and development in the area of electric-arc technology. According to the company, some potential applications for the new alloy are:

- Sensors in water alarms for flood control.
- Temporary inert-gas sealant dams for special weldments.
- Anhydrous-chemical processing.
- Fuel delivery systems in critical diesel installations.
- Encapsulation of organic compounds such as algicides for marine applications.

Series 300 is stable under a wide range of atmospheric conditions showing no sign of erosion or deterioration "over long test periods." However, it disintegrates when immersed in ordinary cold water. (*Materials Engineering*, Mar. 1984, p. 15)

This new alloy was produced as a *byproduct* of other research, and no doubt its dissolving characteristic came as quite a surprise to researchers. The evaluation at the end is, as usual, split into two parts with *However* as the mediating signal. However, the feature of solubility in water, which would normally be regarded as a negative evaluation (a problem), is seen here as possibly a unique property that could make the alloy an ideal solution to something. The manufacturers are making several suggestions in the hope of selling this byproduct as a solution to someone's needs. Here is an example where the *Situation* and *Problem* are known but are not communicated.

6. Urethane Backs up Conductor Traces

A manufacturer of metal on elastomer connectors (MOE) uses Poron cellular urethane as backing material behind conductor traces. Metallic conducting pads are laminated to the resilient, insulating cellular-urethane substrate. Poron urethane provides good compression set resistance and has sufficient resilience to keep matching leads in contact with each other. Results of tests conducted by the resin manufacturer showed that Poron material lost only 10% of its original thickness, when tested for 22 hr. at 158 degrees F(70 degrees C) in accordance with ASTMD3574. (*Materials Engineering*, Mar. 1981, p. 11)

Although the emphasis in this example is on a specific application, the problem the urethane overcomes is not specified, and the situation of electric board manufacture is not given. The writer felt it unnecessary to include this information. Instead he concentrated on describing how the material is used and evaluating it in terms of theoretical consideration of its attributes of set resistance and resilience and also the practical evaluation by the tests. The use of *only* is a signal of good evaluation in this context, although *only* is used as a problem-signaling word or a word of uniqueness in other contexts. The need for study of the meaning of words *in context* is a vital premise of the work described here.

Other Principles

This description is no more than a very brief summary of some of the major principles of rhetorical structure now available in the literature. Other principles also described elsewhere are:

- Detailed discussion of the principles of *Evaluation* in terms of objective basis, subjective assessment and skilled unsubstantiated opinion (Jordan, 1984d).
- Explanation of comparative evaluation with testing, decision, implementation and further testing (Jordan, 1984d).
- Analysis of different types of *Problems* with related signaling words (Jordan, 1984d).
- Discussion of interactional or "people" problems in which an attempted solution by one person (or group) creates a problem for another (Jordan, 1984a).
- Recognition of "condensed" structures in titles and introductory summaries (Jordan, 1981b).

Although this section of the chapter is only a summary of the central features of a theory of rhetorical structures, it is sufficient as a basis for use in practical planning and writing environments, which are now outlined.

Recommendations

Using the Principles

Perhaps the most important use of the principles of rhetorical structures from a pedagogical point of view is as a means of analyzing texts. Analysis should be seen as an important educational aim if the student intends to apply the new-found knowledge to practical writing tasks. This is feasible now that there is a demonstrably sound way to analyze everyday English prose—a form of English that can now be studied to at least the same depth as literary works.

Although the principles of rhetorical structures and signaling have been used for such analytical purposes in language courses in both Canada and England, this anthology is more concerned with their immediately practical applications for practitioners and students of technical writing, and so the remainder of this chapter concentrates on these applications. The methods discussed here have been developed by the author over a number of years in the principal writing course at Queen's University, Kingston,

Canada, "Effective Technical Communication" for senior engineering and science students, and also as instruction to delegates from industry and government on short courses. Three types of application are discussed here: the generation and structuring of material for a report, the writing of a proposal given a situation and problem, and the organization and writing of a report given unstructured material in note form.

Planning a Complex Document

Ironically, the greatest difficulty in writing a complex technical document is not in the actual writing at all—it is in the decision regarding what material to include and how to organize that material. Even more ironic is the sparse guidance available in textbooks on technical writing that provides meaningful help in these important tasks, especially the generation of material to be included. Advice on material generation and selection is now available for technical descriptions (Jordan, 1984c), in which three chapters deal with typical types of information together with organizational patterning partly related to paragraphing. Similar work is needed for nondescriptive texts, and that is the purpose of this subsection.

As a basis for work in an educational environment, we can take any problem-solving situation within the knowledge of the writer and simply ask for an outline detailing both the contents and structure of the proposed document. Some brief introductory outlining instruction and an example or two may first be necessary. One successful topic is based on the following text (discussed in Jordan, 1980).

7. Killer Blanket

> Deaths among elderly people involving electric blankets have increased this Winter so a six-point safety check-list issued by the Minister of State for Prices and Consumer Protection last November is being published again. (*Safety*, Feb. 1979, p. 6)

This complete text is of course a very brief summary of a large amount of material presumably available about this subject. If a writer knows the types of available information, he can select the pertinent information and structure a full report on this subject—even if he has never thought of this subject before. From an earlier discussion in this chapter, we can ask about many aspects of the situation, about the extent and causes of the problem, about the actual solution and other possible solutions, and about their evaluations.

Figure 1.1   A Typical Situation–Problem–Solution–Evaluation Outline

---

TITLE

Abstract

INTRODUCTION

General Material
  Purpose
  Brief overview
  Document structure

Previous Work
  Earlier reports
  Related analyses
  Discussion of standards

Background Material
  Larger elderly population
  Greater need to keep warm
  Increasing costs of general room heating
  Less able in emergencies
  Cause of deaths (fire, electrocution)

PROBLEM

Extent
  Numbers involved
  Comparison (previous years)
  Ages of victims
  Types of home (own or supervised)
  Other (illness, infirmity)

Causes
  Blankets
    Faults (by manufacturer and type)
    Age (wires)
    Instructions (suitable, too complicated?)
  Use
    Overheating due to doubling, extra blankets, etc.
    Malpractice with electrical connections
    Incontinence or leaking hot water bottles
  Other
    Reliance on blanket during electrical outage

SOLUTIONS
  Six-point plan
    Authorship
    Contents (see appendix)
    Circulation
    Differences from last year
  Need to improve standards
  Need to improve instructions
  More need for warnings (clarity, illustrations?)
  Attempted solutions in other jurisdictions

EVALUATION
Possible effect of list last year
Possible effects this year
Need for a permanent "solution"
Final recommendations

APPENDIX
Six-point plan
Related standards
Details of statistics

Other topics could be handled in this way:

1. Acid rain in North America or Europe.
2. Sexist language.
3. Adult illiteracy.
4. Fruit flies.
5. Adult computer illiteracy.
6. Cancer.
7. Pornography.
8. How to use problem-solving structures to plan and write technical documents.

Many others can be used as the basis for the planning exercises here. In practice the writer will be faced with a more constrained problem area to deal with, but the approach learned will help the writer plan and write any complex problem-solving document. The prepared outline (see Figure 1.1) provides a valuable basis for discussion, and it helps the writer to consider all relevant contents and to place these contents in a sensible order with appropriate balance and subdivision of the topics.

Writing a Proposal

Technical writers are often faced with a situation and problem and are asked to propose one or more solutions. The perceived problem can be an inadequacy in a present method or product, a hazardous situation, or a newly defined need. Figure 1.2 illustrates a typical *Problem* statement with several deficiencies, some of which can easily be remedied by a new system of calculation.

Of course, there is no one perfect solution to the defined need here, and in that respect it presents a very realistic task. The writer has to decide which aspects of the deficiencies in the current method warrant some improvement and also which aspects can reasonably be improved. As the "degree day" is purely a means of comparison of the relative coldness for a given period of time or for comparison between years, it could be argued

A MORE USEFUL "DEGREE DAY" DEFINITION

Write a report proposing suitable changes to the "degree day" definition currently used by the Department of the Environment. The report should be suitable for publication in a general interest publication read by engineers and scientists of all disciplines. Here are a few details you will find useful:

1. Current method: take high and low temperature readings between midnight and midnight of the next day, average them, take the average from 18°C (64°F), and you have the "degree day"—a measure of the degree of heating needed for that day.
2. Degree days are added up for the days to arrive at a measure of the amount of heating needed in homes, in offices, etc. during a given period—used by oil delivery companies to assess fuel needs—also a basis of heating efficiency calculations.
3. Because the hours of daylight are fewer than the hours of darkness during winter and temperature highs and lows are related to daytime and nighttime, single readings of high and low are a poor basis for an "average" temperature. What about fast swings of temperature?
4. The method is still the same if the temperature exceeds 18°C for part of the day. Is that reasonable? There can be no negative degree days of course.
5. What about wind? Wind can get in through cracks, and it must also have some effect on insulation effectiveness.
6. What about sunlight? Especially for buildings with large window area, heating by the sun can be significant.
7. 18°C (64°F) is fair enough as an average comfort value during the day, but something less would be OK during the night. Note that offices and schools are not used much at weekends, nor stores on Sundays. Can you take this into account? Are two (or more) degree day values needed?

The present method isn't bad as a rough approximation, and there is no question as to its simplicity, but try to suggest something that you think would work better and justify and describe it. You will also need to explain deficiencies in the existing method and perhaps show how your approach overcomes these. How would you solve problems with your solution, e.g., how would you phase in a new system so that data from previous years can still be used?

Figure 1.2 Typical Proposal Assignment

that the present method is good enough. A report with this point of view would still include the situation, problems and possible solutions, with the evaluation being that the advantages of the new possibilities are outweighed by the simplicity of the present method.

Most writers, however, devise a method that overcomes many of the deficiencies without creating a complicated calculation, although some create a simple empirical formula on a computer program, often including data for hourly temperature, wind, and sunlight. The structure is still *Situation–Problem–Solution–Evaluation*, the final section usually including limitations as well as advantages of the proposed method. A section describing alternative, rejected solutions is often included (with reasons for rejection) when a longer report is requested.

Interestingly, this task shows how closely the thinking process and the writing process are related. The writer has to understand the situation, grasp the problems (and their seriousness and solvability), devise proposed improvements, and consider their relative worth. The readers need this information too—in the same order, separated into suitable sections, and with appropriate signals to indicate the structure and continuity of the proposal. The final sentence of the task statement points out that implementation of any new idea may itself pose a problem. The writer needs to solve this problem and again inform the readers of the difficulty and its resolution.

Arranging and Writing a Report

A third practical circumstance writers face is that they receive information for a report in note form and in no particular order. Their task is first to arrange the material using the structures discussed here and then to write the report including appropriate signals and paragraphing to indicate the structure. An example of this type of practical writing task is given in Figure 1.3, and readers can see an "answer" in *Chartered Mechanical Engineer* (Dec. 1978, p. 25) and Jordan (1980).

For this task the writer needs to have a greater knowledge of rhetorical structures than simply *Situation–Problem–Solution–Evaluation*. Here there is a need to understand that a deficiency with an existing solution is a problem, there can be several competing solutions, these solutions may include some evaluative material, and that actual testing (as opposed to theoretical evaluation) may need to be mentioned. Note also that the "in situ" method of recovering more oil is not a steam drive method and should be seen as almost an appendix separate from the main thrust of the report.

Perhaps the best way to understand the structure of this and similar reports is to illustrate the way the text coheres in some sort of representational array to depict the relationship of different types of information. This has been done for this example in Jordan (1980), which also includes discussion of how the types of information for this example are related to one another in a clause-relational context. In addition, Chapters 5 and 6 of this anthology explain further aspects of coherence and information placement.

Conclusions and Further Applications

These three examples of practical planning and writing tasks illustrate how even a basic knowledge of rhetorical structures helps writers not only to understand the types of information that could, or perhaps should, be present in a given document but also to structure the sentences and paragraphs to

ORGANIZATION AND WRITING WITHIN RHETORICAL STRUCTURES

Rewrite the report "Steam Drives out Oil" using all and only the following information. Put the numbered chunks of information into a suitable information structure first and then write the report. The information presented within each numbered chunk is in the order of the original, but the chunks themselves have to be rearranged.

1. Foster Miller Ltd. to design system burning fuel mixed with air—generates steam by direct contact with water—both steam and combustion gases could be injected further into reservoir—avoids problems of atmospheric pollution.
2. Third contract—World Energy Systems developing down-hole system generator—would produce steam by mixing and burning hydrogen and oxygen at bottom of well.
3. Injecting steam into oil reservoir so far has been limited to wells less than 2000 ft (760 m) deep—reason: at greater depths losses through drill pipe greater—so great steam's efficiency reduced.
4. The DoE plans selection of steam technique with best potential—field testing in 1980.
5. Another Rockwell proposal being studied—use of electric heater at bottom of well to produce steam.
6. Solution to problem may be to generate steam at or near bottom of well—DoE now let three research contracts—purpose: investigate methods of doing this.
7. Used successfully in past—in situ combustion expensive compared with other methods—more efficient than steam driving—applicable to wide range of crude oil—supplies its own fuel—requires only addition of air and water.
8. New methods of extracting more oil from conventional wells by "steam drive" process—to be investigated—to be done by US companies—contracts just let by American Department of Energy.
9. Meanwhile, another DoE contract—General Crude Oil Company investigating another method of recovering additional oil from underground reservoirs—to test in situ combustion process—burns part of the oil in a reservoir—heats remaining oil—reduced viscosity enables combustion gases to drive previously unrecoverable oil to producing well.
10. Rocketdyne Division of Rockwell International working on similar idea—steam generated in "down-hole" heat exchangers—not by direct contact—exhaust gases vented to atmosphere—avoids possibility of "plugging" the reservoir with particles generated during fuel combustion.

Figure 1.3 Sample Writing Task

communicate the contents most effectively. Within and between different types of information, writers need to include appropriate linguistic signaling devices to help readers to readily comprehend the information and the relationship between the types of information. The use of a large number of such signaling devices is discussed in detail and indexed in Jordan (1984d).

These three examples illustrate three methods of using this work. Other applications, however, have also been useful, and the following list provides some wider perspective of the possible uses of this material in technical writing:

1. Case studies in which details of the situation and problem (e.g., a critical injury) are provided in conversational form, with the writer being asked to write the report including situation and problem, with detailed analysis of the causes of the problem and with suggested remedial action in the form of recommendations.
2. Exercises involving editing to improve conciseness, grammar, punctuation, etc. that are also examples of rhetorical structure patterns and that need improvement in linguistic signaling, paragraphing, and arrangement of material.
3. Presentation of a report that is poorly arranged with respect to communication of two or more problems or two or more solutions, with the writer having the task of rearranging the material and making necessary changes to bring out the newly arranged structure more clearly— this sort of work is ideal for use on a word processor.
4. Open-ended assignments in which the writer selects a situation and problem(s) and then devises and communicates appropriate solutions.

In addition, for those involved with teaching the structures of English prose, selected examples can be analyzed by students (singly or in groups) and the written analysis presented. Many examples are analyzed in this way in Jordan (1984d), which has the aim of helping students and practitioners to understand and then analyze structured texts.

Final Comments

When used with knowledge and sensitivity, the rhetorical patterns described here can be applied to the overall structure of a very large number of technical reports, articles, and proposals. In addition, the principles can be applied to many parts of documents, as minor problem-solving thought/action patterns are often found within larger structures. Finally, the thinking behind the selection, sequence and communication of relevant material is a valuable basis for the writing of summaries and precis. In total, this chapter demonstrates that there is a powerful unspoken linguistic consensus regarding thought/action structures in many technical tasks. This consensus is a useful tool in the selection of relevant material, its arrangement, and its writing.

This chapter is devoted to the important pedagogical problem of accounting usefully for the sequence and contextual meanings of sentences within texts. These contextual meanings represent certain definable types and subtypes of high-priority information that readers will expect to be included in the text. These types of information are recognized by readers by the sequencing and by the use of special linguistic indicators within the text. The study of rhetorical structures should therefore not deal only with types of information and their sequencing; it should also involve analysis of the linguistic signals that signpost the structures and thus help readers to follow the pattern of the discourse. Similarly for writing, practitioners need to be aware of the power of the structure-signaling vocabulary within the work they are creating.

The discerning reader will detect an even wider level of significance. Although this chapter concentrates on analyzing, planning, and writing technical texts, those following the widely held consensus described here will also be following the thought/action patterns of engineers and scientists as they do their work of creation or discovery. The thought/action patterns and the related rhetorical patterns in texts are essentially the same; the purpose of the text is to describe what happened and why and what the result was. Study of the organization and signaling of structured technical texts should therefore be seen to have a wider pedagogical significance than simply helping the student or technical writer improve his writing; it should also be seen as an effective and teachable means of enabling the student or technical writer to grasp the essence of problem-solving and evaluation, which are the fundamental aspects of any professional education.

Although the patterns and signaling described here account for the overall structure of a very large number of technical texts, readers should realize that there are texts that may not be structured in this way because they are essentially "descriptive" in nature. Descriptive technical writing relies almost entirely on cohesion between items of the vocabulary to ensure the thread of continuity in the text (as described in Jordan, 1982b, for example). Nevertheless even purely descriptive texts follow a *Solution* structure as any purposeful engineering project clearly seeks a solution. In addition many technical descriptions contain evaluative statements, as the writer indicates how good (or, occasionally, how bad) the product is; and any information can be taken as the basis for the reader to make her/his own assessment of the worth of the product or service being described. Thus descriptive texts are special cases of the consensus described here.

Other structures commonly involved in technical texts are those we can broadly class as logical. The relations of logic have been discussed in general terms by a number of linguists, and a useful classification of the systems involved is now available (Jordan, 1985c) with some elaboration in Jordan (1984b). In these works, logic is divided into three reciprocal pairs of relations (cause–effect, purpose–means, and basis–assessment), all of which overlap in many cases with the structure described in this chapter. Clearly, any problem-solving activity contains many assessments—each with its own basis. It also involves actions taken to meet a desired purpose and results of such actions. Exactly how these different relations work in English texts, together with complicating relations of fruition (Was it done?), cognition (Do you know?), and modal semantics (propriety, willingness, desire, etc.), remains to be determined. However, it seems likely that eventually an overall theory of rhetorical structure can be described that explains both the structure and signaling of texts and the thought/action patterns of the events being described in all logically determined problem-solving activities. The absence of discussion here on the relations of logic is deliberate and necessary given the present state of knowledge of the subject.

Structures involving comparison (included in Jordan, 1984e, 1988b) can also be seen to be part of the overall structure described here. Any comparison involves both description and implicit evaluation. Other relations (such as those of affirmation and denial and those of surprise and expectation) may eventually be seen as relations mapped over the basic structures to add additional information supplementary to the main message of the text (Jordan, 1984f).

References

Beekman, J., & Callow, J., (1974). *Translating the word of God*. Grand Rapids, MI: Zondevan Publishing House.

Christophersen, P. (1939). *The articles: A study of their theory and use in English*. Copenhagen: E. Munksgaard.

Earl, J. H. (1977). *Engineering design graphics*. Reading, MA: Addison-Wesley.

Edel, D. H. (1967). *Introduction to creative design*. Englewood Cliffs, NJ: Prentice-Hall.

Grimes, J. (1975). *The thread of discourse*. The Hague: Mouton Press.

Gross, B. J. & Sidner, C. L. (1986). Attention, intentions and the structure of discourse, *Computational Linguistics*, **3**, 175–204.

Halliday, M. A. K. (1967). Notes on transitivity and theme in English, *Journal of Linguistics*, **3**, (Part 1) 37–81; (Part 2) 199–244; **4**, (Part 2) 199–244; **4**, (Part 3) 179–215.

Halliday, M. A. K., & Hasan, R. (1976). *Cohesion in English*. New York: Longman.

Hawkins, J. A. (1978). *Definiteness and indefiniteness*. London: Croon-Helm; Atcaetic Highlands, NJ: Humanities Press.

Hoey, M. P. (1979). Signalling in discourse, *Discourse Analysis Monograph No. 6*. University of Birmingham, England.

Hoey, M. P. (1981). Discourse centered stylistics: A way forward? In W. Gutwinski & G. Joly (Eds.), *LACUS Forum* (pp. 401–409). Columbia, SC: Hornbeam Press.

Hoey, M. P. (1983). *On the surface of discourse*. Winchester, MA: Allen & Unwin.

Huddleston, R. D. (1971). *The sentence in written English*. Cambridge, MA: Cambridge University Press.

Huddleston, R. D., Hudson, R. A., Winter, E. O., & Henrici, A. (1968). *Sentence and clause in scientific English*. London: University of London Press.

Jordan, M. P. (1980). Short texts to explain problem-solution structures and vice versa. *Instructional Science, 9*, 221–252.

Jordan, M. P. (1981a). Some associated nominals in technical writing. *Journal of Technical Writing and Communication, 11*, 3, 251–264.

Jordan, M. P. (1981b). Structure, meaning and information structures of some very short texts. In W. Gutwinski & G. Joly (Eds.), *LACUS Forum* (pp. 410–417). Columbia, SC: Hornbeam Press.

Jordan, M. P. (1982a). Fundamentals of technical description. In *The Proceedings of the 29th International Technical Communications Conference* (p. 64). Boston, MA: Society for Technical Communication.

Jordan, M. P. (1982b). The thread of continuity in functional writing. *Journal of Business Communication, 19*, 4, 5–12.

Jordan, M. P. (1982c). Structured information in functional writing. *Teaching English in the Two-Year College, 9*, 61–64.

Jordan, M. P. (1983). Complex lexical cohesion in the English clause and sentence. In A. Manning, P. Martin, & K. McCalla (Eds.), *LACUS Forum* (pp. 224–234). Columbia, SC: Hornbeam Press.

Jordan, M. P. (1984a). Structure, style, and word choice in everyday English texts. *TESL Talk, 15*, 1–2.

Jordan, M. P. (1984b). Some clause relational associated nominals in English. *Technostyle, 4*(1), 36–46.

Jordan, M. P. (1984c). *Fundamentals of technical description*. Melbourne, FL: Krieger Press.

Jordan, M. P. (1984d). *The rhetoric of everyday English texts*. Winchester, MA: Allen & Unwin.

Jordan, M. P. (1984e). Co-associative lexical cohesion in technical publicity. *Journal of Technical Writing and Communication, 16*(½), 33–55.

Jordan, M. P. (1984f). Some relations of surprise and expectation in English. In R. A. Hall, Jr. (Ed.), *LACUS Forum* (pp. 263–273). Columbia, SC: Hornbeam Press.

Jordan, M. P. (1985a). Close cohesion with *do so:* A linguistic experiment using a multi-example corpus. In B. Couture (Ed)., *Functional approaches to writing: research perspectives* (pp. 29–48). Wakefield, NH: Francis Pinter.

Jordan, M. P. (1985b). Non-thematic re-entry: An introduction to and extension of the system of nominal group reference/substitution in everyday English use. In J. D. Benson & W. S. Greaves (Eds.), *Systemic perspectives on discourse*, Vol. 1 (pp. 322–332). Norwood, NJ: Ablex.

Jordan, M. P. (1985c). Systems of logic in technical writing. *Technostyle, 4*(3), 1–6.

Jordan, M. P. (1986a). Applying language systems to the teaching of writing. *The Technical Writing Teacher, 13*(3), 212–225.

Jordan, M. P. (1986b). Teaching writing based on computer-based language research. Paper presented at the Canadian Association for Teachers of Technical Writing. Montreal, Quebec, Canada. *Technostyle,* 5(2), 11–17.

Jordan, M. P. (1988a). Advances in clause stationed theory. In J. D. Benson & W. S. Greaves (Eds.). *Systemic Functional Approaches to Discourse* (pp. 282–301). Norwood, NJ: Ablex Press.

Jordan, M. P. (1988b). *Structure in English Texts.* New York: The Edwin Mellen Press.

Lyons, J. (1968). *Introduction to theoretical linguistics.* Cambridge, MA: Cambridge University Press.

Mann, W. C., & Thompson, S. A. (1983). *Relational proposition in discourse.* Report ISI/RR-83-115. Los Angeles: University of Southern California, Instructional Sciences Institute.

Mann, W. C., & Thompson, S. A. (1987). Rhetorical structure theory: A theory of text organization. Report ISI/RS-87-190. Los Angeles: University of Southern California, Instructional Science Institute.

Winter, E. O. (1971). *Connection in science material: A proposition about the semantics of clause relations.* Report No. 7. London: Centre for Information on Language Teaching.

Winter, E. O. (1976). *Fundamentals of information structure.* Unpublished manuscript. The Hatfield Polytechnic Institute, Hatfield, Hertfordshire, England.

Winter, E. O. (1977). A clause-relational approach to English texts: A study of some predictive lexical items in written discourse. *Instructional Science,* 6(1), Special issue.

2

How Can Technical Writers Write Effectively for Several Audiences at Once?

V. MELISSA HOLLAND, VEDA R. CHARROW, and
WILLIAM W. WRIGHT

V. Melissa Holland, research psychologist for the Army Research Institute, Veda R. Charrow, writing consultant, and William W. Wright, writing specialist at the Document Design Center, pose a typical yet challenging problem for writers. How do writers prepare a technical document for several audiences all of whom will use the document for different purposes? Holland, Charrow, and Wright argue that successfully completing a document with apparently contradictory objectives depends on the necessary but difficult understanding of the document's disparate audiences and manipulating the organization and writing style in the document accordingly. Writers must carefully define the various readers, identify the factors that make these readers different, and anticipate how adjustments can be made in the organization and format of a document so that all the designated audiences have a fair chance of understanding and, more importantly, using the document.

Holland, Charrow, and Wright draw on relevant research in cognitive theory and psycholinguistics and on their own experience to suggest exactly how writers can address several audiences at once. As the authors point out, recent research clarifies many distinctions among potential readers. For example, readers approach documents with acquired patterns for understanding and organizing information; their background and prior knowledge can affect their comprehension dramatically; their familiarity with a subject and reading ability can be measured and categorized; and all of these variables can be addressed in concrete ways to improve the usefulness of functional documents. Furthermore, research confirms that readers' goals will dictate most major decisions about a document's overall organization and format; these goals can generally be classified according to the readers' professional positions as managers, researchers, scientists, or technicians. Writers who

27

understand the different expectations of their readers can manipulate a text to fulfill various audiences' needs by using, for example, complex or simple prose, flowcharts or discussion, formulas or explanations.

The authors suggest ways of reaching readers, from the conventional attention to word and sentence length to concentration on more unusual schemes and presentations such as multimedia combinations, diagramming, graphical highlighting, and story-like translations. Finally, the authors recommend detailed descriptions, case studies, and methods for overcoming the major obstacles presented by multiple audiences. Using their professional experience, Holland, Charrow, and Wright also discuss how financial and political considerations influence a document's design.

Multiple Audiences

Composing text for several audiences at once is a task familiar to most technical writers—for example, writing a report in which the methods will be analyzed by technicians and the conclusions considered by policymakers. But it is a task that logically and practically poses complex problems.

Logically, writing for multiple audiences requires higher-level strategies than writing for one audience. Communicating effectively with one audience is itself complex, as the other chapters in this book demonstrate, because it requires, for example, deciding what information is new to the reader, integrating that information with what the reader already knows, and expressing the information clearly and directly. But communicating with multiple audiences is even more difficult. For example, since the features that make a text *more* effective for one audience may make it *less* effective for another audience, writers must weigh the trade-offs involved in using one or another set of features.

Practically, there is little published guidance for dealing with several audiences at once. Most writing guidelines look at how to deal with a single general audience or with one specific type of audience, such as the content expert or the reader who lacks technical vocabulary.

Thus, writers addressing a mixed audience must rely on intuition and experience. When these are inadequate, the resulting text can be conspicuously ineffective. Readers with less technical knowledge than those to whom the text is geared may not understand it. Readers interested only in parts of the text may waste time finding what they want to know. In either case, the intended message may never reach one set of readers. At best, one audience is served at the expense of another.

In this chapter we make recommendations about how to write for multiple audiences. Because the base of empirical research on this problem is as slim as the base of published practical guidance, many of our recommendations come from unpublished sources—from our own experience in a center devoted to designing technical documents and training document designers and from case studies of how other technical writers have successfully handled the problem.

In the next section of this chapter, A Framework for Defining the Problem, we outline a way to specify your writing task in particular instances. Who are your different audiences? In what ways and to what extent do they differ? Your answers to these questions will guide you in solving a multiple-audience problem. Also included is a way to specify the constraints of time and cost—since these will influence how you solve the problem. In outlining this framework, we cite a variety of research to support our recommendations about what kinds of differences to pay attention to in writing for diverse readers.

In the third section, Relevant Research, we describe ways to deal with each of the major types of multiple-audience problems identified in our framework. In the Recommendations section, we draw on different sources to support the solutions we describe: the existing research and practical literature on the problem, our own experience and expertise, and a collection of case studies we gathered in developing this chapter. These studies are representative of the kinds of problems you may have in writing for a mixed audience, and they illustrate effective solutions. They appear in the last section of this chapter, A Set of Case Studies.

Our framework of problems and solutions applies to two basic kinds of technical materials: expository prose (reports, proposals, memos, functional descriptions) and instructions (directions, procedures, manuals). Fill-in-the-blank forms and prose that is primarily narrative or persuasive are not directly dealt with in our framework.

A Framework for Defining the Problem

There is no blanket solution to the problem of writing for multiple audiences. Solutions depend on the dimensions of your particular problem. The following questions will help you define the critical dimensions:

- In what ways and to what extent do the anticipated audiences for your text differ?
- What are your constraints in producing the text?
- Is one audience more important than another?

Let us consider the range of answers you might have for these questions and the kinds of answers that matter most for technical writing.

In What Ways and to What Extent Do Your Audiences Differ?

Understanding the ways and extent your audiences differ will allow you to determine the nature of the problem and the solution. Although there are no fixed rules for deciding how much an audience has to vary to affect what you write, you can generally look for clear, categorical differences in any of the following dimensions:

- The audience's *familiarity* with the topic. (Are your readers subject–matter experts or novices?)
- The reading *ability* of the audience. (Are your readers college level or are they adults reading at an 8th-grade level?)
- The audience's *goals* in reading. (Do your readers need to analyze the technical details of a study to make policy based on the study's conclusions?)
- The audience's reading *tasks*. (Do your readers want to read and learn or to use the text as a reference?)

An audience may vary in numerous other ways, such as cognitive style, motivation, or preference for words versus pictures. But those sorts of differences are probably not powerful enough to threaten the overall effectiveness of a text designed for one kind of reader and not another. Here we are concerned only with the dimensions for which audience variation is likely to have strong effects on the comprehensibility and usability of your writing. (See Jonassen, 1982, for a review of selected reader variables and their effects.)

Relevant Research

What research leads us to assume that the audience dimensions listed above will have strong effects? Let us look more closely at each dimension.

Familiarity with the Topic

A reader's familiarity with a topic is a matter of specific exposure, not of intelligence or ability. This dimension has been investigated in research on reading in a variety of fields—cognitive science, education, and psycholinguistics. The research confirms our intuitions: readers who know something about a topic find it easier to grasp new material about the topic than readers who lack relevant prior knowledge. The more prior knowledge readers have, the faster they read the material, the better they understand it, and the more they remember (Anderson, 1981; Anderson et al.,

1977; Anderson, Spiro, & Anderson, 1978; Chiesi, Spilich, & Voss, 1979; Clifton & Sloviaczek, 1981; Garner, 1979; Graesser, Hoffman, & Clark, 1980; Johnson & Kieras, 1982).

Depending on the material to be read, prior knowledge can make dramatic differences. For example, readers who know something about a topic before they read a document will remember two and three times more information than uninformed readers (Bransford & Johnson, 1973; Bransford & McCarrell, 1974; Dooling & Mullet, 1973). Additionally, readers with diverse kinds of prior knowledge have been found to interpret passages in different and conflicting ways (Anderson et al., 1977; Pichert & Anderson, 1977). Such effects occur particularly when the experimental passages are opaque or potentially ambiguous—as is frequently the case with technical material.

In explaining these effects, cognitive theory assumes the people acquire highly organized, detailed knowledge about the world—sometimes called "schemas"—which they use to make sense of incoming information (Adams & Collins, 1979; Bransford, 1979; Minsky, 1975; Rumelhart & Ortony, 1977; Schank & Abelson, 1977; Spiro, 1977). Reading is an interaction between these mental schemas and the words on the page. Without applicable schemas, readers have little basis for interpreting what they read (Anderson, 1977; Graesser, Hoffman, & Clark, 1980).

Moreover, research suggests ways to help readers who lack the right schemas. Their comprehension can be vastly improved by providing a background. This background can be in the form of orienting directions (Pichert & Anderson, 1977), explanatory statements and titles (Dooling & Mullet, 1973), diagrams or pictures of the events being referred to (Bransford & Johnson, 1973), or introductory material that helps readers acquire the knowledge needed to understand a document (Anderson, 1981).

Research and theory thus indicate that readers' familiarity with the topic is an essential variable for technical writers to contend with. Clearly, then, an audience of readers who *differ* in their familiarity raises special problems. Research identifies three levels of familiarity as significant in readers: those readers who don't know the topic (novices); those who know the topic well (experts); and those who are technically informed but not expert. For example, if the topic is how to use a new word-processing system, novice readers might be those who have never used a computer, experts would be specialists in word-processing systems, and technically informed readers would be those who have occasionally used word-processing software.

In addition, research identifies as significant readers those who are experts in different fields related to the topic. For example, if the topic is developments in medical thermography, one aduience might be doctors using medical thermography equipment whereas the other might be equipment engineers.

Differences such as these should be clear in a preliminary audience analysis or in the specifications you receive from a contractor or client. If these sources do not define the levels of technical knowledge, then it is a good idea to seek more information about your readers' backgrounds. If you discover a group of nonspecialists in your audience, be sure to find out what terms they are likely to need defined (see Mathes & Stevenson, 1976).

Reading Ability

Numerous studies confirm the importance of reading ability in text comprehension. The lower a reader's measured reading skill, the less the reader understands and remembers from given passages (e.g., Kern, 1979; Kern et al., 1976; Kniffen et al., 1980). Moreover, the practical impact of this variable is well documented in settings where the skills of adult readers vary widely: studies of military training (Kniffen et al., 1980; Muckovak, 1979; Sticht, 1979; Sticht & Zapf, 1976) and of social welfare programs (Bendick & Cantu, 1978; Charrow et al., 1980; Holland & Redish, 1982a). These studies show that mismatching the skills of readers and the skill levels required by documents is pervasive and carries high operational costs.

Research and practice have explored what the reader's ability implies for how best to structure a text. For low-ability readers, practical precedent typically calls for using shorter words and sentences (Bendick & Cantu, 1978; Caylor et al., 1973; Kincaid et al., 1975; Klare, 1979). Beyond these measures, practical precedents include adopting familiar formats (like personal letters), translating terms into story-like definitions, and using situations and examples (Charrow et al., 1980; Holland, 1981).

In addition, experimental work has shown that low-ability readers frequently do better when the structure of written passages is made highly explicit. This can mean (1) breaking the text into small chunks (Jonassen, 1982), (2) using headings, adjunct questions, maps placed before the text, and other "advance organizers" (Dean & Kulhavey, 1979; Duchastel, 1982; Mayer, 1979; Rothkopf, 1972), or (3) making the relations between sentences explicit with cohesive words, graphical highlighting, pointers (like "note"), and preview and summary statements (Duchastel, 1982; Golinkoff, 1976; Meyer, 1977, 1980; Neilsen, 1978; Pace, 1980). Low-ability readers also seem to do better with alternatives to text: oral or audiotaped presentations (Gray, Snowman, & Deichman, 1977; Samuels & Horowitz, 1980) or pictures and diagrams (Holliday, Brunner, & Donais, 1977; Stone, Hutson & Fortune, 1984). These methods of simplifying materials are generally recommended for readers less literate than the average high school graduate.

An audience that *ranges* in reading abilities warrants particular concern. If some part of your audience is expected to read below the 12th-grade level, then this should be clear from your audience analysis or the specifications from your client. If reading test scores are unavailable, you can estimate literacy levels from education levels (Bendick & Cantu, 1978). In addition, you can generally expect non-native English speakers who are new to the language to read English less well than native English speakers—as found in military and other special populations (Holland & Redish, 1982b; Holland, Rosenbaum, & Stoddart, 1982).

Goals of Reading

Depending on their goals, readers may want different kinds of information from a text. This is common among readers of technical material, who frequently have different jobs involving different concerns. The executive may want only the summary and conclusions of a report evaluating a new telecommunications system. This information might help the executive decide whether to adopt the system for the company. The resident engineer may need the technical sections of the report, including the primary statistical results, the functional explanations, and the procedures used to collect and analyze data. That information would help in understanding and implementing the telecommunications system.

Distinctions in readers' goals may coincide with their levels of prior knowledge. The executive is likely to be uninformed about the technical aspects of telecommunications, whereas the engineer knows details. However, readers with different levels of knowledge may have uniform reading goals, just as readers with different goals may have equivalent knowledge. For example, one technician may read equipment brochures to decide whether to purchase the equipment, another to find out how to maintain it.

Very little experimental work explores what part readers' goals play in determining how comprehensible or useful a text will be. The applied literature, however, stresses the importance of understanding readers' goals in order to write effectively (e.g., Dodge, 1962; Houp & Pearsall, 1980; Joseph, 1979; Lannon, 1979; Mathes & Stevenson, 1976; Pearsall & Cunningham, 1978). The same literature documents the costs of using texts ill-matched to readers' goals. When material is structured around one purpose, it confuses or slows down those who read it for another purpose.

The differences in reading goals that merit the most concern are those with major consequences for selecting and organizing the content of the text. The applied literature classifies these goals in terms of jobs that characterize people who use technical information: (1) managers or executives, (2) researchers or scientists, and (3) technicians, engineers, or equipment users. (See Houp & Pearsall, 1980; Lannon, 1979; and for military jobs, Kern et al., 1976.) Executives are usually concerned with conclusions, interpretations, and recommendations and with information about marketability, costs and benefits, risks, and impacts. Researchers are more concerned with raw data, theoretical calculations, and scientific explanation. Technicians or equipment users want step-by-step procedures and practical details about assembly, maintenance, or operation.

The consequences of these goals for selecting and sequencing text content are clear. The applied literature advises that the content of most concern to each type of audience be presented first or exclusively. Thus, executive reports should give conclusions and recommendations first (Houp & Pearsall, 1980; Joseph, 1979), whereas research reports give these last (Pearsall & Cunningham, 1978). Moreover, evidence from surveys and psychological experiments suggests that these opposing sequences match the expectations of the executive and the researcher respectively (Dodge, 1962; Kintsch & Van Dijk, 1978).

Reading Tasks

The canonical reading task is straight-through, serial processing of a text; the readers' task in serial reading is to learn and remember information. However, readers also engage in other kinds of tasks, especially with technical material. Two important kinds of reading tasks are *reference reading,* to find an answer, and *performance reading,* to carry out a set of procedures.

The readers' task may be influenced by, but is not identical to, the reader's goal. By goal we mean the general kind of information the readers want to extract. By task we mean the process of extraction itself. Thus, even though their goals may differ, both the executive and the technician can use a text either for serial reading or for looking up answers to specific questions. Similarly, a technician can use a set of procedural instructions to learn the procedure or to perform it without learning.

A variety of literature supports the importance of the task dimension in the reading process. Research on reading suggests that underlying the three tasks we have distinguished are distinct psychological processes (Coke, 1976; Duffy & Kabance, 1982; Graesser, Hoffman, & Clark, 1980; Pugh, 1978; Sticht, 1979; Wright & Reid, 1973). Reference reading requires scanning and search processes; serial reading requires memory processes; and performance reading requires processes of planning and translating into actions.

These processes also appear to have major implications for how best to structure a text. For example, simplifying the language matters little for reference reading but significantly for serial reading (Coke, 1976; Duffy & Kabance, 1982). Also, prose is the optimal format to present rules for the readers to learn. But flowcharts and other forms of algorithms are better when the readers merely use the rules to answer a question or perform a procedure (Holland & Rose, 1981; Kammann, 1975; Wason, 1968; Wright & Reid, 1973). This is because logorithms can be used quickly and automatically, without any need to understand the underlying rule.

Furthermore, the prose format that people expect for serial reading and learning is expository paragraphs, but for performance it is numbered steps (Gordon et al., 1978). The critical organizational feature to aid learning of expository paragraphs is hierarchical structure—for example, initial summary or goal statements (Johnson & Kieras, 1982; Kieras, 1981; Kintsch & Van Dijk, 1978; Meyer, 1977, 1980). But the critical feature to aid performance of procedures is the ease of translating individual steps into actions (Dixon, 1982; Kieras, 1984; Smith & Spoehr, 1984). However, if procedures are to be learned and remembered, then hierarchical structure appears to be essential (Smith & Goodman, 1982). Hierarchical goals and purpose statements aid memory for procedures by chunking together and explaining individual steps.

The applied literature follows the research by urging writers to organize and format differently when the text is used for reference, serial reading, or how-to performance (Kern et al., 1976; Mathes & Stevenson, 1976; Pearsall & Cunningham, 1978). For example, texts serving serial reading and learning can be organized by topics and concepts, whereas texts serving procedural performance are to be organized by tasks (see Black, 1984; Kern et al., 1976; Sticht, 1979). In addition, the literature advises that reference texts be given a more highly structured, extensively cross-referenced organization, with wide use of headings, captions, indices, and other aids to search and retrieve information (*Simply Stated*, 1984; Waller, 1982, 1983). Flowcharts and other job aids are recommended for reference and performance tasks (Gane, Horabin, & Lewis, 1966; Kern et al., 1976; Wheatley & Unwin, 1972).

Thus, differences in the readers' task warrant different text organizations and formats. Clearly, readers who will do more than one task with a given text require special measures.

What Are Your Constraints in Producing the Text?

A variety of external constraints influence the kinds of solutions you can use in dealing with multiple audiences. The most influential kinds of constraints include:

- Time.
- Money and staffing.
- Contextual limitations on page size, text length, use of color, type of language, etc.

Time and money constraints can affect your planning and writing activities as well as the way the manuscript is produced. Time and money limitations can reduce the extent of the audience analysis you do, the graphics' possibilities you consider (like pictures and multiple color), and the overall alternatives you have (such as creating separate documents for separate audiences).

Contextual constraints are imposed by politics, your client's preferences, and the conditions under which the text will be used. Contextual constraints might include the requirement of pocket-sized reference cards, training circulars that are light enough to carry into the field, or monochromatic graphics for conventional journals. In addition, political considerations might dictate the use of professional jargon rather than simple language, as shown in one of our case studies (Case Study 2).

Is One Audience More Important Than Another?

How you deal with multiple audiences also depends on whether one audience is more important than another. For example, one of our case studies (3) illustrates a criminal justice program proposal that is addressed to program administrators and program implementors. Because administrators are responsible for deciding whether to fund programs, the author decided to make them the primary audience. Thus, *political expediency* can determine whether you have a primary audience and how to satisfy that audience.

Another example involves legal writers in the U.S. government who publish descriptions of investments that involve pension plan funds. Since the writers' primary purpose is to explain what is happening to people's money, they consider pension plan participants to be their first audience. But since lawyers sometimes also read the descriptions to find out the legal implications of various investments, the writers consider lawyers a secondary audience. Thus, the writer's *purpose* and *relative size* of the audiences can dictate the primary reader.

Recommendations

If your audience varies in any of the dimensions listed in the previous section, how can you make your writing as understandable and usable as possible to all readers? In this section we describe ways to deal with each major type of mixed audience.

Some of the research and applied literature cited earlier describes ways to accommodate an audience that fits a single distinct category. For example, you can start with conclusions and recommendations when your readers are executives; you can use picture supplements for nonspecialists. Solutions of this sort are widely accepted and have good support in the research and practical literature.

There is little literature on how to accommodate an audience with several categories of readers. However, there are real-world precedents: many writers have handled combined audiences with apparent success.

Some of these precedents are briefly mentioned in a few sets of guidelines for technical writing (Houp & Pearsall, 1980, pp. 92–93; Huckin, 1983, p. 102; Kern et al., 1976, pp. 21). The most extensive treatment is given by Mathes and Stevenson (1976, pp. 12–23). We draw on these guidelines in developing the solutions described here. We also draw on our own experience in document design and on the experience of several professional writers whom we interviewed for the case studies that conclude this chapter. Finally, we draw on some of the single-audience solutions described in the first section and show how these can be incorporated into multi-audience solutions.

Arriving at solutions is ultimately a creative process. Our framework of problems and solutions can be viewed as an aid to this process, offering dimensions to consider and concrete examples of solutions that have worked.

A Pool of Techniques for Treating Multiple Audiences

We first summarize the techniques most commonly recommended for dealing with multiple audiences. Many of these techniques apply to more than one type of multiple audience:

1. Writing *different documents*—manuals, brochures, reference cards, or circulars—for each audience (Kern et al., 1976).
2. Directing the language and presentation of a *single text* to the lowest level of reading skill or the lowest level of technical understanding found in the target audiences (Jonassen, 1982; Lannon, 1979).
3. *Compartmentalizing* a single text so that separate sections are read by

separate audiences (Houp & Pearsall, 1980; Mathes & Stevenson, 1976). Compartmentalization can involve:

- Using "access structures" to help readers find different information in a text. These structures include (1) indexes (*Simply Stated,* 1984; Waller, 1982), (2) tables of contents, outlines, and preliminary summaries and objectives (Waller, 1982); (3) headings, captions, page tabs that mark sections, and advance organizers (Duchastel, 1982; Houp & Pearsall, 1980; Jonassen, 1982);
- Building in several optional modes of presentation (e.g., text and pictures) and levels of detail side by side in a single text or program. This allows readers to select the level or mode they need as they read (Black, 1984; Horn, 1982; Kieras, Tibbits, & Bovair, 1984; Stone, Hutson, & Fortune, 1984);
- Placing definitions, explanations, or technical details in footnotes, glossaries, appendixes, or special sections within the text (Houp & Pearsall, 1980; Pearsall & Cunningham, 1978).

The third solution, compartmentalizing a single text, can be applied to nearly every type of mixed audience we discuss. If you compartmentalize your text, consider the following general guidelines and cautions. First, label the separate sections you have created with precise and informative headings so that readers can easily find what they need. By informative headings we mean those that relate directly to questions the reader is likely to have, rather than those that name a general topic. For example, a heading like "How to apply for financial assistance" is more useful than "financial assistance." (See also Horn, 1982; Swarts, Flower & Hayes, 1980.) A title like "What is a pension benefits plan?" introduces background sections better than "Pension benefits plans." Informative headings are helpful for a single audience because they provide handles for understanding and remembering (Charrow & Redish, 1980; Mayer, 1979; Rothkopf, 1972; Swarts, Flower & Hayes, 1980). But they are critical for a multiple audience, since they tell what sections to read and what to skip.

Second, identify clearly in the introduction and the table of contents what and where the different sections are (see Felker & Rose, 1981, and Case Study 7). If some of these sections are appendixes or glossaries, call attention to them early in the text, at the points where readers are likely to need them. One recommended way to direct readers explicitly to the right sections is by using branching directions (Charrow et al., 1980; Waller, 1983). These directions can go in an introduction ("If you are new to word processing, read Section 3 first") or at the end of individual sections ("If you need to read more about the keyboard, go to Appendix A").

Third, be cautious in using appendixes. They make reading easier for those who don't want or need the material, but relegating too much information to an appendix strips the text of meaning, leaving it cryptic and sometimes incomprehensible (see Houp & Pearsall, 1980; Joseph, 1979). Appendixes should be used for optional information, and they should be well organized in themselves—for example, different appendixes for different purposes and material.

Finally, use a thorough index for reference or instructional manuals. This helps readers find exactly where to turn (see Case Study 4).

Techniques for Treating Specific Kinds of Multiple Audiences

How do the techniques we have summarized relate to the four kinds of multiple audiences described earlier? What other techniques can we suggest? Our recommendations appear below. In some of our examples, the audiences differ in more than one dimension. If you encounter audiences like this, you may want to look at more than one set of recommendations.

Audiences That Differ in Their Familiarity with the Topic

To convey the same basic information to an audience with different levels of knowledge about the topic, it is usually better to use a *single text*. You can then present one set of pictures, diagrams, conclusions, or other material likely to be understood by all readers. For instructional manuals, an advantage of the single text is that readers can continue to use it as they become more advanced. This is important, since different users may become expert at different times on different sets of instructions.

Separate texts make sense if your readers' levels of knowledge are widely disparate. For example, reporting on experimental research for a scientific journal and for a popular magazine generally calls for two unique manuscripts. Frequently, differences in knowledge combine with differences in the basic tasks readers are expected to do. Here too, separate texts may work better (see "Reading Tasks" in this section).

There are two ways to accommodate a single text to the needs of specialists and nonspecialists: (1) compartmentalize it or (2) write it for the nonspecialist. If you compartmentalize, you can (a) attach separate sections to the same core material or (b) present separate versions side by side with no overlapping core.

Attaching Separate Sections

You can *compartmentalize* a single text so that the core information is read by everyone and designated sections are read by only the specialists or only the nonspecialists. (Case Studies 3 and 8 are illustrations of this.) This solution is particularly useful for reports and proposals. For the nonspecialist, the sections appropriate to separate out are background information and definitions. For the specialist, the appropriate sections are the technical details. These sections can be presented in introductions, footnotes, or text supplements like glossaries and appendixes. The advantages of each kind of presentation are discussed below—first for the nonspecialist and then for the specialist.

If the *nonspecialist* needs background knowledge to understand the main body of a report or a procedure, you can provide this background either in an introduction or in an appendix. The choice between an introduction and an appendix depends in part on whether one audience is primary and another is secondary. For example, if topic experts are your primary readers and nonspecialists are secondary, then background information would be better in an appendix.

If the nonspecialist needs help with terminology, and the background section does not define all the terms (or you do not need a background section), you can define terms outside the text. Use footnotes to the body of the text or a glossary at the end of the text. Footnotes and glossaries each involve trade-offs. Footnotes are easier for the reader than a glossary because they don't require flipping back and forth between pages in the text. On the other hand, for ease of publication, a glossary is more practical, especially if you anticipate changes and updates. Changes can be made to the glossary without disturbing the page layout in the body of the text. Footnotes usually work better for serial reading; glossaries for reference reading. In serial texts, the first occurrence of a technical term can be footnoted. In reference texts, it is impossible to know which of several occurrences of a term a reader will see first.

One modern format for presenting footnote information is a highlighted box. This box can be outlined or shaded in a different color from the rest of the page, then placed in a margin or side column so as not to interrupt the flow of text. Textbooks and procedural manuals often use boxes throughout the text to present definitions, caveats, or examples. By consistently formatting certain types of information this way, you help readers to process the text selectively. For example, some readers may want to isolate definitions and return to them often, while others may want to skip them entirely.

If the *specialist* is likely to need technical details that the nonspecialist won't understand, these can go outside the text in appendixes and footnotes or in the text in specially marked chapters or sections (see Case Study 5). Footnotes suit short technical comments; appendixes suit longer sections—for example, statistical tables, protocol transcripts, or citations of legal precedent (see Houp & Pearsall, 1980; Pearsall & Cunningham, 1978). Whether to remove technical matter from the text or leave it, marked, in the text depends in part on how important experts are as an audience. In the example referred to earlier, government legal writers consider lawyers a secondary audience for descriptions of pension fund investments. These writers use footnotes to subordinate information about the legal implications of particular investments, information meaningful mainly to lawyers. The footnotes also contain more technical language than the text. In Case Study 5, the mathematical formulas are in footnotes for the convenience of both the nonmathematical audience and the mathematical audience.

Presenting Separate Versions

You can *compartmentalize* a text by presenting *separate versions* with no overlapping core. This solution is more appropriate for procedures and user manuals (where some readers are likely to be more advanced than others) than it is for reports or proposals. You can present more and less advanced versions of the same material in two ways: (1) in separate places in the text or (2) side by side in the text.

Presenting more and less advanced versions in separate places is illustrated by Case Study 1. Here the writer built a computer manual around detailed explanations of the steps required to do computer tasks. At the end of the manual, the writer attached a "quick reference card," an overview of the steps without explanation, designed for the more experienced user. Another variation on this is Case Study 7, in which the writers used a very detailed table of contents as a summary of the text.

Presenting more and less advanced versions side by side is illustrated by Case Study 4. Here the writer built "fast paths" through a computer user manual. One step in building fast paths is to list instructional summaries in the margins beside each set of detailed instructions. The advanced user can read these summaries and avoid the detail. Another example of this technique is structured writing or information mapping (Horn, 1982). In computer-presented instructions, simultaneous options have proven effective in studies by Kieras, Tibbits, and Bovair (1984) and Stone, Hutson, and Fortune (1984). Kieras, Tibbits, and Bovair present a hierarchical menu of instructions for operating an electronic device. The expert could stop at "Set the electrometer to 50"; the novice could select the detailed steps that tell how to set the electrometer. Stone, Hutson, and Fortune present basic assembly instructors that readers could supplement with pictures or definitions by pressing a computer key.

Finally, Case Study 1 illustrates a form of self-paced instruction to accommodate more or less advanced computer users. Here, instead of choosing their own options, users are directed by the program or text to the appropriate place in a sequence of lessons. The direction depends on how the user performed on the previous lesson.

Your resources, or the brevity of the text, may not permit compartmentalization. In that case, we recommend writing for the general reader (see Case Study 3 and Case Study 6, Part I). You can still use technical terms, but use them sparingly and, preferably, define them in the body of the text. Since your fundamental concern is to convey information, it is better to risk condescension to the expert than nonsense to the novice. (Refer to the section "Familiarity" for other ways to help the novice.)

Audiences That Differ in Their Reading Abilities

If reading ability is the sole difference between groups of readers, then the most efficient solution is a *single text* written for the low-ability group. In general, the various techniques that help low-ability readers will not hurt average or high-ability readers. Thus, you can employ simplified language (shorter sentences, more common words), explicit text structure (summary statements, signaling devices, cohesive words), short paragraphs, advance organizers (questions and headings), story-like examples, and picture supplements without impairing the good reader's ability to use the document. There is some evidence that good readers *prefer* unsimplified materials when they are asked to choose among different versions (Holland, 1982; Morris, Thilman, & Myers, 1979). However, preference in one group is usually secondary to comprehension in all groups, especially when you are dealing with highly functional, nonpersuasive material.

You can also *compartmentalize* the text by offering simultaneous picture–text options, as discussed under "Familiarity." Stone, Hutson, and Fortune (1984) found that readers from lower-ability groups tended to choose picture options more than did higher-ability readers. Having the choice enabled both groups to perform instructions accurately and with minimum delay.

Audiences That Differ in Their Reading Goals

If readers are expected to vary in the kind of information they seek from a document, then *compartmentalizing* the document (Houp & Pearsall, 1980; Mathes & Stevenson, 1976) and creating *separate documents* (Kern et al., 1976) are both commonly recommended solutions. Your choice between these solutions will depend in part on available resources: it is less expensive to publish and distribute a single document than separate documents. Your choice will also depend on the length of the text: when the compiled segments add up to an unwieldy manuscript, you probably need separate documents. When these two factors are of equal weight, consider how the information is to be used.

Compartmentalizing is useful when the same reader wants to be aware of all parts of the document or when closely related readers use the document in tandem (see Case Study 8). Thus, in a proposal to adopt a new heat exchange system, an initial segment for the executive could provide a cost analysis or other data relevant to decision making, plus conclusions and recommendations. A second segment for technicians could contain functional explanations, installation and maintenance procedures, and a description of the techniques used to analyze and compare heat exchange systems. The executive could skim over these parts and assign sections of interest to technical experts. The initial segment would be addressed to executives because they are the primary audience. In addition, the appendixes and footnotes can also help to set apart information for readers with different goals.

Separate documents are useful when the respective audiences don't need to refer to each other's material. Thus, in a report on last year's trade law reform project, high-level managers may not want to deal with a bulky manuscript and probably have little need to refer to technical sections. You could write a lengthy technical report and put a succinct executive summary under separate cover, as in Case Study 6, Part II. For example, major U.S. trade associations submit detailed technical memos on industrial trade issues to government lawyers and economists and send brief issue statements to the chief executive officers of member industries.

Separate documents are also useful in training manuals intended for different job groups in the military. Kern et al. (1976) list several disadvantages for a single, partitioned manual: it is bulky and inconvenient for transportation and field use; it usually has a consolidated index, which takes longer to use than a separate index; it is generally less accessible and less useful to each of the groups that deals with it.

Audiences That Differ in Their Reading Tasks

Single, *compartmentalized texts* are usually appropriate when some readers use the text for reference and some for serial reading, or when some use the text to learn a procedure and some to perform it without learning. Single texts help especially if the same readers are expected to do more than one of these tasks. For instance, serial readers may also have questions to look up. But if task differences combine with other audience differences, then separate texts may work better.

For example, a *single* text can serve one group of readers learning about a subject or a system and another group with selected questions about the subject or system. The text could be designed for serial reading but enriched by numerous informative headings and page tabs—to facilitate skimming and searching for information—and a highly developed index—to facilitate cross-referencing (see Case Study 4). However, *separate texts* for serial reading and for reference may be better if the audience reading serially is also very new to the subject area and the audience that is referring questions is experienced. For example, many computer companies design a learning primer for computer novices, organized sequentially from simple to complex, and a separate reference manual, organized by tasks, for informed users. This solution avoids overwhelming the novice.

A *single* text can serve one group of readers learning about a system and another group carrying out procedures on the system. For example, naval manuals are divided into parts for functional explanation and parts for troubleshooting and maintenance procedures (Duffy, 1982). If you use this method, refer to "Reading Tasks" for specific ways to adapt the respective parts to serial reading and to performance reading. In general, use expository prose for serial reading, hierarchically structured to facilitate understanding and learning. The prose can be accompanied by illustrations, models, and other diagrams that aid understanding. Use step-by-step instructions for procedural performance, accompanied by flowcharts or other algorithmic formats that can be used to perform tasks without knowing the rules behind the tasks. These formats will speed performance. But if step-by-step procedures are to be learned and remembered, embed them in hierarchies where goal and purpose statements explain the separate steps (for example, "These next steps will help you remove the hubcap so you can get to the wheel").

A Set of Case Studies

Case Study 1   Writer A, a technical writer, was writing an instructional manual that shows a new user how to begin using a personal computer. Writer A faced the problem of having to write for the "computer naive" person as well as for the person who may have had experience on several other systems.

Writer A felt that the best way to deal with the problem of multiple audiences was to present the instructions as a series of exercises geared toward the less experienced user—but to build in features to allow the more sophisticated user to move at a faster pace. At the end of each exercise, the user is given the opportunity to sign off the computer or go on to a more challenging lesson. Writer A also created a quick reference card (a summary of the main commands needed to operate the computer). By using this card, attached to the back of the manual, the user who had experience on other systems could skip some of the elementary step-by-step instructions and begin operating the computer.

Case Study 2   Writer B was a technical writer who had also written advertising copy and had developed campaign literature for people running for governor and U.S. senator in an eastern state with a diverse population. In discussing ways to write for multiple audiences, she was able to draw on years of experience that went beyond writing technical instructions. She felt that to write effectively for multiple audiences, the writer should always begin by getting a detailed profile of the audience. If there is a large disparity between audiences, the writer should write two different pieces.

In writing for state political campaigns, this writer broke her audience into "clusters." These included groups such as coal miners, young urban professionals, and the elderly. Then she and other campaign officials looked closely at the purpose of the information she needed to get across and how different the groups or clusters were. At this point she made decisions about whether she could write one piece of material to appeal to several groups or whether several pieces were necessary. Writer B stressed that when writing advertising copy that must appeal to several audiences, writing campaign literature, or writing technical instructions, the important thing is to get accurate information about the audience (or audiences) and be sure you understand the purpose of the piece of writing.

Case Study 3   Writer C had written a manual for a national criminal justice project. The manual was developed with the help of experienced criminal justice managers from around the country and had several purposes and several audiences. One purpose was to encourage criminal justice administrators to set up citizen participation programs at their institutions; another purpose was to suggest ways to manage these programs more effectively. The audience included upper-level decision makers at a variety of juvenile and criminal justice institutions (from prisons to juvenile detention homes) as well as management-level people who would implement the procedures.

The committee of experts responsible for developing the manual wished to appeal to the decision maker or major policymaker first (in order to get the volunteer program implemented) and the manager second (in order to keep the program running efficiently). Some members of the committee wanted the manual to be written in the jargon of the primary audience—the criminal justice administrator—rather than in plain English. They felt that simple prose might be a good way to present instructions for the manager (the secondary audience) but that it might offend the agency administrator whose attention they wanted to grab first. They felt that to persuade they needed to write in the language (or jargon) of the primary audience. Writer C convinced his committee of experts that the manual would both persuade and instruct effectively if written in a style that all audiences could understand. However, he did compartmentalize the manual. Sections aimed specifically at the policymaker or manager carefully explained who must do what. The manual addressed the primary audience (the policymakers) first, and every chapter heading was written so that the reader knew what the section contained and could avoid it if it did not apply to him or her.

The problem Writer C encountered highlights another problem we all have probably encountered when writing technical materials—that is, appealing to a hidden audience, the person (or group of people) who commissioned the document. The person who writes the writer's paycheck might have a different idea of who the writer is writing for. Fortunately, Writer C was able to convince his review committee that it was not necessary to use jargon to appeal to the policymakers in the criminal justice system.

Case Study 4 Writer D, who edits and manages the production of manuals for a large computer company, solves the problem of multiple audiences by presenting materials for naive and sophisticated readers side by side in the text. He builds "fast paths" into the manuals he develops so that the more advanced user can move through them more quickly. A "fast path" is any graphic or textual device that allows people to move through a text at the speed that suits them best. Some "fast paths" he uses are:

- Headings that are informative.
- Comments in the margins.
- Icons or visual cues for certain functions (for example, color-coded sections for advanced readers).
- Instructions in the margin for the experienced user (not as detailed as the step-by-step procedures for the beginner).
- Tabs that inform the reader what sections contain.
- A good index.

Interestingly, Writer D, a strong advocate of "plain English," pointed out that an index with only plain English terms can be useful for the new user but can frustrate the person who knows only the jargon of a certain field. A new user might look for "start"; an older one might look for "boot."

Case Study 5 Writer E, a lawyer/mathematician, wrote a law review article in which he used mathematical probability theory to shed light upon a legal issue. Writer E knew that the major audience for the article (lawyers) would not have the mathematical sophistication to understand the formulas he was using. However, he also knew that the article would be scrutinized carefully by the small group of mathematicians and mathematician/lawyers who are interested in applications of probability theory to the law.

He solved the problem by writing the article for the nonmathematician, explaining probability theory and how it applied to the legal issue. He also explained his mathematical proofs in words. The proofs themselves were in footnotes, so that the mathematically sophisticated reader could follow the mathematical argument. In this way, the purely legal audience did not have to stumble over the math and the mathematical audience had merely to go from footnote to footnote to follow the mathematical logic.

Case Study 6 Writer F, an experienced rewriter of technical and legal documents, has used a number of different techniques to deal with multiple audiences.

I. In some cases, she writes to the most important audience, gearing organization, sentence structure, and vocabulary to that audience. This means that if the most important audience consists of unsophisticated readers, other audiences may find the document too simple. If the most important audience consists of specialists in a field, less sophisticated readers may have some difficulty understanding it. Nonetheless, the most important audience will be satisfied with the document.

II. In other cases, when the constraint on one audience is lack of reading time, Writer F writes different pieces for the different audiences. For example, for a government report that was meant to be read by technicians and by senior managers, she wrote the report for the technicians but included a three-page executive summary for the managers. The executive summary distilled the important points of the report and presented them in a short and coherent fashion that managers could use as a decision-making tool.

Case Study 7 Writers G and H, a lawyer and an economist at the Federal Trade Commission (FTC), conducted a lengthy investigation of the true costs of life insurance. The Staff Report they wrote at the end of their investigation was intended for the commissioners of the FTC, lawyers in other government agencies, and for any members of the public and the insurance industry who might be interested in the findings from the investigation. The potential problem for the audiences of this document was not its technical difficulty: the intended audiences consist of people who could reasonably be expected to understand the material. Rather, the potential problem was the length of the document: the report itself is 183

TABLE OF CONTENTS

Figure 2.1 A Table of Contents Used as a Quick Synopsis of a Report, Illustrating a Form of Execution Summary

pages long and the appendixes add another 200 pages or so. Some subsections of the audience—commissioners and decision makers within the agency, busy executives outside the agency—might not have the time to read the entire report. Ordinarily, an executive summary might solve this problem. But the authors of this report came up with an ingenious alternative. They used informative headings throughout the paper, much like the issue statements and answers that serve as headings in a legal brief. Gathered into a table of contents at the front of the book, these headings serve as a summary of the entire study (see Figure 2.1). The busy reader has only to read the table of contents to understand the study, its findings, its conclusions, and its recommendations. He or she also has the option of reading or skimming a specific section of the report to get more information.

Case Study 8 Writers I and J of another agency solved the problem of multiple audiences in an equally creative way. In writing a regulation for the Health Education Assistance Loan (HEAL), these writers at the Department of Health, Education, and Welfare realized that there were several audiences:

PART 126—HEALTH EDUCATION ASSISTANCE LOAN PROGRAM

Subpart A—General Program Description

Sec.
126.1 What is the HEAL program?

Subpart B—The Borrower

126.5 Who is an eligible student borrower?
126.6 Who is an eligible nonstudent borrower?
126.7 The loan application process.
126.8 What are the borrower's major rights and responsibilities?

Subpart C—The Loan

126.10 How much can be borrowed?
126.11 Terms of repayment.
126.12 Deferment.
126.13 Interest.
126.14 The insurance premium.
126.15 Other charges to the borrower.
126.16 Power of attorney.
126.17 Security and endorsement.
126.18 Consolidation of HEAL loans.
126.19 Forms.
126.20 The Commissioner's collection efforts after payment of a default claim.
126.21 Refunds.

Subpart D—The Lender

126.30 Which organizations are eligible to apply to be HEAL lenders?
126.31 The application to be a HEAL lender.
126.32 The HEAL lender insurance contract.
126.33 Making a HEAL loan.
126.34 HEAL loan servicing.
126.35 HEAL loan collection.
126.36 Consequence of using an agent.
126.37 Forbearance.
126.38 Assignment of a HEAL loan.
126.39 Death and disability claims.
126.40 Procedures for filing claims.
126.41 Determination of amount of loss on claims.
126.42 Records, reports and inspection requirements for HEAL lenders.
126.43 Limitation, suspension, or termination of the eligibility of a HEAL lender.

Subpart E—The School

126.50 Which schools are eligible to be HEAL schools?
126.51 The student loan application.
126.52 The student's loan check.
126.53 Notification to lender of change in enrollment status.
126.54 Payment of refunds by schools.
126.55 Administrative and fiscal procedures.
126.56 Records.
126.57 Reports.
126.58 Federal access to school records.
126.59 Records and Federal access after a school is no longer a HEAL school.
126.60 School audits.

AUTHORITY: Title VII, part C, subpart I of the Public Health Service Act, as amended, 90 Stat. 2243 (42 U.S.C. 294–294*l*).

Subpart A—General Program Description

§126.1 What is the HEAL program?

(a) The health education assistance loan (HEAL) program is a program of Federal insurance of educational loans designed for students in the fields of medicine, osteopathic medicine, dentistry, veterinary medicine, optometry, podiatry, pharmacy, and public health. The basic purpose of the program is to encourage lenders to make loans to students in these fields who desire to borrow money to pay for their educational costs. In addition, certain nonstudents (such as doctors serving as interns or residents) can borrow in order to pay the current interest charges accruing on earlier HEAL loans.

(b) Schools, State agencies, pension funds, banks and other financial institutions having an insurance contract with the Commissioner of Education can be lenders under the HEAL program. HEAL lender eligibility is described in §126.30.

Figure 2.2 Organization by Audience Needs (*Federal Register,* Vol. 43, No. 150, August 3, 1978)

- The borrower—the student who is seeking the loan.
- The lender—the bank or other institution that will be accepting the student and administering the loan.

Accordingly, the writers broke the regulation into sections, each addressed to the particular audience (see Figure 2.2). In this way, any audience could avoid technical information or instructions that were irrelevant to that audience, and still have enough information to act on. The one general section on the loan itself could be read by all audiences.

References

Adams, M. J., & Collins, A. (1979). A schema-theoretic view of reading. In R. O. Freedle (Ed.), *New directions in discourse processing,* Vol. 2 (pp. 1–22). Norwood, NJ: Ablex.

Anderson, J. R. (1981). Effects of prior knowledge on memory for new information. *Memory and Cognition, 9,* 237–246.

Anderson, R. C. (1977). The notion of schemata and the educational enterprise: General discussion of the conference. In R. C. Anderson, R. J. Spiro, & W. E. Montague (Eds.), *Schooling and the acquisition of knowledge* (pp. 415–431). Hillsdale, NJ: Erlbaum.

Anderson, R. C., Reynolds, R. E., Schallert, D. L., & Goetz, E. T. (1977). Frameworks for comprehending discourse. *American Educational Research Journal, 14,* 367–381.

Anderson, R. C., Spiro, R., & Anderson, M. C. (1978). Schemata as scaffolding for the representation of information in connected discourse. *American Educational Research Journal, 15,* 433–440.

Bendick, M., & Cantu, M. G. (1978). *The literacy of welfare clients.* Washington, DC: The Urban Institute. (Reprinted from *Social Science Review, 52.*)

Black, J. (1984). *Helping people learn to use computer systems.* Paper presented at the Document Design Center, Washington, DC.

Bransford, J. D. (1979). *Human cognition: Learning, understanding, and remembering.* Belmont, CA: Wadsworth Press.

Bransford, J. D., & Johnson, M. K. (1973). Considerations of some problems of comprehension. In W. G. Chase (Ed.), *Visual information processing* (pp. 383–438). New York: Academic Press.

Bransford, J. D., & McCarrell, N. (1974). A sketch of a cognitive approach to comprehension: Some thoughts about understanding what it means to comprehend. In W. Weimer & D. Palermo (Eds.), *Cognition and the symbolic processes* (pp. 189–279). Hillsdale, NJ: Erlbaum.

Caylor, J. S., Sticht, T. G., Fox, L. C., & Ford, J. P. (1973). *Methodologies for determining reading requirements of military occupational specialities.* HumRRO Technical Report 73–5. Presidio of Monterey, CA: Human Resources Research Organization.

Charrow, V. R., Holland, V. M., Peck, D. G., & Shelton, L. V. (1980). *Revising a*

Medicaid recertification form: A case study in the document design process.
Washington, DC: Document Design Center.

Charrow, V. R., & Redish, J. C. (1980). *A study of standardized headings for warranties.* Technical Report 6. Washington, DC: Document Design Centers.

Chiesi, H. L., Spilich, G. J., & Voss, J. F. (1979). Acquisition of domain-related information in relation to high and low domain knowledge. *Journal of Verbal Learning and Verbal Behavior, 18,* 257–273.

Clifton, C., & Slowiaczek, M. L. (1981). Integrating new information with old knowledge, *Memory & Cognition, 9,* 142–148.

Coke, E. U. (1976). Reading rate, readability, and variations in task-induced processing. *Journal of Educational Psychology, 68,* 167–173.

Dean, R. S., & Kulhavy, R. W. (1979). *The influences of spatial organization in prose learning.* Paper presented at the Annual Meeting of the American Educational Research Association, San Francisco, April.

Dixon, P. (1982). Plans and written directions for complex tasks. *Journal of Verbal Learning and Verbal Behavior, 21,* 70–84.

Dodge, R. W. (1962). What to report. *Westinghouse Engineer, 22,* 108–111.

Dooling, D. J., & Mullet, R. L. (1973). Locus of thematic effects in retention of prose. *Journal of Experimental Psychology, 97,* 404–406.

Duchastel, P. C. (1982). Textual display techniques. In D. H. Jonassen (Ed.), *The technology of text,* Vol. 1 (pp. 167–191). Englewood Cliffs, NJ: Educational Technology Publications.

Duffy, T. M. (1982). *Preparing technical manuals: Specifications and guidelines.* Presented at American Educational Research Association, New York.

Duffy, T. M., & Kabance, P. (1982). Testing a readable writing approach to text revision. *Journal of Educational Psychology, 74,* 733–748.

Felker, D., & Rose, A. M. (1981). *The evaluation of a public document: The case of FCC's Marine radio rules for recreational boaters.* Washington, DC: Document Design Center.

Gane, G. P., Horabin, I. S., & Lewis, B. N. (1966). The simplification and avoidance of instruction. *Industrial Training International,* I. Report No. 4, 160–166.

Garner, R. (1979). *The importance of cognitive styles research for understanding of the reading process.* Paper presented at the Annual Meeting of the American Educational Research Association, San Francisco (ED 172 148), April.

Golinkoff, R. M. (1976). A comparison of reading comprehension processes in good and poor readers. *Reading Research Quarterly, 11,* 623–659.

Gordon, L., Munro, A., Rigney, J. W., & Lutz, K. A. (1978). *Summaries and recalls for three types of text.* Technical Report 85. Los Angeles: University of Southern California.

Graesser, A. C., Hoffman, N. L., & Clark, L. F. (1980). Structural components of reading time. *Journal of Verbal Learning and Verbal Behavior, 19,* 135–151.

Gray, L. R., Snowman, J., & Deichman, J. (1977). *The effect of stimulus presentation mode and cognitive style on sentence recognition memory.* Paper presented at the Annual Meeting of the American Educational Research Association, New York.

Holland, V. M. (1981). *Psycholingistic alternatives for readability formulas.* Technical Report 12. Washington, DC: Document Design Center.

Holland, V. M. (1982). *Applying text design guidelines to medical consent forms.* Paper presented at American Educational Research Association, New York.

Holland, V. M., & Rose, A. M. (1981). *A comparison of prose and algorithms for presenting complex instructions.* Technical Report 17. Washington, DC: Document Design Center.

Holland, V. M., & Redish, J. C. (1982a). Strategies for understanding forms and other public documents. In D. Tannen (Ed.), *Proceedings of the Georgetown Round Table on Discourse: Text and Talk* (pp. 205–218). Washington, DC: Georgetown University Press.

Holland, V. M., & Redish, J. C. (1982b). *An audience assessment of MOS/11B language skills.* BSEP interim report. Washington, DC: American Institutes for Research.

Holland, V. M., Rosenbaum, H., & Stoddart, S. C. (1982). *BSEPI/ESL programs: Findings and program descriptions* (technical report). Washington, DC: American Institutes for Research.

Holliday, W. G., Brunner, L. C., & Donais, E. L. (1977). Differential cognitive and affective responses to flow diagrams. *Journal of Research in Science Teaching, 14,* 129–138.

Horn, R. E. (1982). Structured writing and text design. In D. H. Jonassen (Ed.), *The technology of text,* Vol. 1 (pp. 341–367). Englewood Cliffs, NJ: Educational Technology Publications.

Houp, W., & Pearsall, T. E. (1980). *Reporting technical information.* Beverly Hills, CA: Glencoe.

Huckin, T. N. (1983). A cognition approach to readability. In P. V. Anderson, R. J. Brockmann, & C. R. Miller (Eds.), *New essays in technical and scientific communication: Research theory and practice* (pp. 90–108). Farmingdale, NY: Baywood.

Johnson, W., & Kieras, D. (1982). *The role of prior knowledge in the comprehension of simple technical prose.* UARZ/DP/TR-82/ONR-11. Tuscon: University of Arizona.

Jonassen, D. H. (1982). Individual differences and learning from text. In D. H. Jonassen (Ed.), *The technology of text,* Vol. 1 (pp. 444–463). Englewood Cliffs, NJ: Educational Technology Publications.

Joseph, A. M. (1979). *Professional and technical writing.* Detroit, MI: General Motors Education & Training Management Development Curriculum.

Kammann, R. (1975). The comprehensibility of printed instructions and the flowchart alternative. *Human Factors, 17,* 90–113.

Kern, R. P. (1979). *Usefulness of readability formulas for achieving Army readability objectives: Research and state-of-the-art applied to the Army's problem.* Fort Benjamin Harrison, IN: Technical Advisory Service, U.S. Army Research Institute.

Kern, R. P., Sticht, T. G., Welty, D., & Hauke, R. N. (1976). *Guidebook for the development of army training literature.* Washington, DC: U.S. Army Research Institute for the Behavioral and Social Sciences, HumRRO.

Kieras, D. E. (1979). *The role of global topics and sentence topics in the construction of passage macrostructure.* Technical Report 6. Tucson: University of Arizona.

Kieras, D. E. (1981). Topicalization effects in cued recall of technical prose. *Memory & Cognition, 1,* 541–549.

Kieras, D. E. (1984). *Reading in order to operate equipment.* Paper presented at the American Educational Research Association, New Orleans, LA.

Kieras, D. E., Tibbets, M., & Bovair, S. (1984). *How experts and non-experts operate electonica equipment from instructions.* Technical Report 14. Tucson: University of Arizona.

Kincaid, J. P., Fishburne, R. P., Rogers, R. L., & Chissom, B. S. (1975). *Derivation of new readability formulas (Automated Readability Index, Fog Count, and Flesch Reading Ease Formula) for Navy enlisted personnel.* Research Branch Report 8-75. Millington, TN: Naval Air Station Memphis.

Kintsch, W., & Van Dijk, T. A. (1978). Toward a model of text comprehension and production. *Psychological Review, 85,* 363–394.

Klare, G. R. (1979). *Readability standards for army-wide publications.* Evaluation Report 79-1. Fort Benjamin Harrison, IN: U.S. Army Administrative Center.

Kniffen, J. D., Stevenson, C. R., Klare, G. R., Entin, E. B., Slaughter, S. L., & Hooke, L. (1980). *Operational consequences of literacy gap.* AFHRL-TR-79-22. Brooks Air Force Base, TX.

Lannon, J. M. (1979). *Technical writing.* Boston: Little, Brown.

Mathes, J. C., & Stevenson, D. W. (1976). *Designing technical reports: Writing for audiences in organizations,* Indianapolis, IN: Bobbs-Merrill.

Mayer, R. E. (1979). Can advance organizers influence meaningful learning? *Review of Educational Research, 49,* 371–383.

Meyer, B. J. F. (1977). The structure of prose: Effects on learning and memory and implications for educational practice. In R. C. Anderson, R. J. Spiro, & W. E. Montague (Eds.), *Schooling and the acquisition of knowledge* (pp. 179–200). Hillsdale, NJ: Erlbaum.

Meyer, B. J. F. (1980). *Text structure and its use in the study of reading comprehension across the adult life span.* Paper presented at the Annual Meeting of the American Educational Research Association, Boston, April 7–11.

Minsky, M. (1975). A framework for presenting knowledge. In P. H. Winston (Ed.), *The psychology of computer vision* (pp. 211–277). New York: McGraw-Hill.

Morris, L. A., Thilman, D. G., & Myers, A. M. (1979). *Application of the readability concept to patient-oriented drug information.* Rockville, MD: U.S. Food and Drug Administration.

Muckovak, W. P. (1979). *Literacy skills and requirements of Air Force career ladders.* AFHRL-TR-74-90, AD-A008-770. Lowry Air Force Base, CO.

Neilsen, A. R. (1978). *The role of text structure on the comprehension of familiar and unfamiliar written discourse.* Paper presented at the National Reading Conference, St. Petersburg, FL.

Pace, A. J. (1980). *The ability of young children to correct comprehension errors: An aspect of comprehension monitoring.* Paper presented at the Annual Meeting of the American Educational Research Association, Boston.

Pearsall, T. E., & Cunningham, D. H. (1978). *How to write for the world of work.* New York: Holt, Rinehart & Winston.

Pichert, J. W., & Anderson, R. C. (1977). Taking different perspectives on a story. *Journal of Educational Psychology, 69,* 309–315.

Pugh, A. K. (1978). *Silent reading: An introduction to its study and teaching.* London: Heinemann.

Rothkopf, E. Z. (1972). Variable adjunct question schedules, interpersonal interaction, and incidental learning from written material. *Journal of Educational Psychology, 63*, 87–92.

Rumelhart, D. E., & Ortony, A. (1977). The representation of knowledge in memory. In R. C. Anderson, R. J. Spiro, & W. E. Montague (Eds.), *Schooling and the acquisition of knowledge* (pp. 99–135). Hillsdale, NJ: Erlbaum.

Samuels, S. J., & Horowitz, R. (1980). *Good and poor reader recall of oral and written expository discourse at two levels of difficulty.* Paper presented at the Annual Meeting of the American Educational Research Association, Boston (ED 138).

Schank, R. C., & Abelson, R. P. (1977). *Scripts, plans, goals, and understanding: An inquiry into human knowledge structures.* Hillsdale, NJ: Erlbaum.

Simply Stated (1984). The neglected index. Document Design Center *Newsletter, 44*, 1–3.

Smith, E. E., & Goodman, L. (1982). *Understanding instructions: The role of explanatory material.* TR-5088. Cambridge, MA: Bolt Beranek and Newman, Inc.

Smith, E. E., & Spoehr, K. T. (1984). *Understanding and executing instructions.* Paper presented at Office of Naval Research Contractors' Meeting, University of Arizona.

Spiro, R. J. (1977). Remembering information from text: The "state of schema" approach. In R. C. Anderson, R. J. Spiro, & W. E. Montague (Eds.), *Schooling and the acquisition of knowledge* (pp. 137–165) Hillsdale, NJ: Erlbaum.

Sticht, T. G. (1979). Developing literacy and learning strategies in organizational settings. In H. F. O'Neil & C. D. Spielberger (Eds.), *Cognitive and affective learning strategies* (pp. 275–307). New York: Academic Press.

Sticht, T. G., & Zapf, D. W. (Eds.) (1976). *Reading and readability research in the armed services.* HumRRO Tech. Rep. 76-A. Alexandria, VA: Human Resources Research Organization.

Stone, D. E., Hutson, B. A., & Fortune, J. C. (1984). *Using the computer to study on-line information selection process.* Paper presented at the American Education Research Association, New Orleans, LA.

Swarts, H., Flower, L. S., & Hayes, J. R. (1980). *How headlines in documents can mislead readers.* Technical Report 9. Washington, DC: Document Design Center.

Waller, R. (1982). Text as diagram: Using typography to improve access and understanding. In D. H. Jonassen (Ed.) *The technology of text,* Vol. 1 (pp. 137–166). Englewood Cliffs, NJ: Educational Technology Publications.

Waller, R. (1983). Designing a government form: A case study. *Information Design Journal, 4*(1), pp. 3–21.

Wright, P., & Reid, F. (1973). Written information: Some alternatives to prose for expressing the outcomes of complex contingencies. *Journal of Applied Psychology, 57*, 160–166.

Wason, P. C. (1968). The drafting of rules. *The New Law Journal, 118*, 548–549.

Wheatley, P., & Unwin, D. (1972). *The algorithm writer's guide.* London: Longman.

3

How Can Technical Writing Be Persuasive?

SCOTT P. SANDERS

For language is framed to convey not the object alone, but likewise the character, mood, and intentions of the person representing it.
Samuel Taylor Coleridge, *Biographia Literaria* (1817)

Most writers of functional documents ignore rhetoric when they write because they perceive it as, at best, a showy and ornate use of language or, at worst, a deliberately deceptive use of language. Rhetorician and professional writer Scott P. Sanders discusses the errors of such opinions by tracing rhetoric's origins, major concerns, and contemporary applications in technical writing. As Sanders explains, technical writers have long argued that rhetorical principles do not apply to functional documents because their subject matter, typically science, reflects objective truth. Thus, technical writers would suggest, the most effective functional prose is that which promotes data, the empirical bases of science, and not opinion and persuasion, the supposed domain of rhetoric.

However, it is commonly argued that the concept of objectivity in science is itself not objective. Rather than a collection of truths, scientific facts are a set of assertions judged objective or worthwhile because a community agrees on their general application and importance. When new data are introduced, conscientious scientists rethink their generalities, often changing a perceived objective truth for another position that explains more of the observed phenomena.

Just as scientists redefine objective truth by rearranging their theories, technical writers create objectivity in their reports not by the reports' subject matter but by effectively arranging the elements of the document's structure to meet the needs of its readers. Therefore, Sanders argues, if writers manipulate the structure of a text to ensure objectivity and make the document persuasive, they engage in the rhetoric they wish to avoid. Sanders condemns the common practice of altering the words of a functional document or its sentences by applying readability formulas or embracing the trappings of an alleged "plain style." Rhetoric used properly is persuasive because it has as its goal a form of mediation, a mutual understanding by writers and readers of the content of a document.

Sanders recommends that technical writers use Rogerian argument. In its succinct form, Rogerian argument progresses from a writer's neutral statement of a problem to a discussion of that problem in the context of the audience's interests, to the writer's evaluation and recommendations to the audience. In a larger context, this argumentative form is a persuasive tactic that presents information in technical documents in an objective, compelling manner.

Persuasion as Rhetoric

The commonplace assertion in technical writing has been that a technical writer has only one valid persuasive strategy: to write so clearly that the facts of a document speak for themselves and convince readers to understand and then to act on their understanding quite apart from the writer's personal motive. This assertion is grounded in a truth. Clear writing is persuasive; and, granting the often complex or obscure nature of its content, the first imperative of technical writing must be clear exposition.

For this reason, most rhetorical analyses of technical writing discuss expository structures designed to present facts clearly and coherently; persuasiveness is taken for granted as an implicit by-product of effective exposition. This is unfortunate. Rhetoric defined as the clear exposition of facts alone neglects the reader's affective responses. It is often not effective enough to ensure the reader's understanding, much less the reader's assent and subsequent action.

Most technical writers would immediately accept that persuasive rhetoric plays a valid role in proposal writing. But even in proposal writing, the use of persuasion is limited. Specialists in proposal writing claim that "it is all but impossible to win with a poor proposal," implying that contracts may be won or lost because of the persuasive qualities of the proposal document itself apart from the specific content of the proposition (Holtz, 1979, p. ix). But when they discuss "technical persuasion" (Whalen, 1982, p. 33), they recall standard expository advice: present "factual data, given in a logical framework" (p. 33) that aims "to express, not to impress" (p. 35); that is, proposal writers should appeal preeminently to the reader's intellect through clear, coherent exposition. Rhetoricians have argued since antiquity that the most effective expository writing complements and augments its intellectual content with a consciously controlled, affective impact on the reader.

Why has there been, until only very recently, so little research on persuasive strategies in technical writing? Today's commonplace understanding of the word "rhetoric" to mean fancy, but empty, talk provides a clue. This usage is the residue of a misunderstanding and subsequent distrust of rhetoric that probably goes back to Aristotle's definition: "Rhetoric may be defined as the faculty of observing in any given case the available means of persuasion" (Barnes, 1984, p. 2155). Taking this statement out of its context, some readers have understood Aristotle's willingness to consider "in any given case the available means of persuasion" to show slight regard for truth. In fact, nothing could be further from the truth (see Lunsford, 1979, p. 150). But the mistaken identification of rhetoric with verbal trickery and deceit remains strong. And perhaps no group of professional writers has rejected rhetoric more loudly, consistently, yet incorrectly than have technical writers.

Relevant Research

The Rejection of Rhetoric

Technical writing developed as a distinct subgenre of what English prose historians call "the plain style," a mode of writing that came into being during the 17th century as writers responded to calls for stylistic reform issued by many diverse groups (Jones, 1930). Among the reformers were the members of the Royal Society of London who decried rhetorical excesses and prescribed a neutral, objective style as the proper method for reporting the results of scientific experiments. Writing in 1667, Bishop Thomas Sprat, in his *History of the Royal Society* (1958, p. 113), described the society's commitment to plain prose as "a constant Resolution to reject all amplification, digressions, and swellings of style; to return to the primitive purity and shortness, when men deliver'd so many things almost in an equal number of words . . . bringing all things as near the Mathematical plainness as they can. . . ." This passage is commonly cited as one of the earliest and most telling indications of the origins of technical writing as a distinct genre (see Halloran, 1978; Halloran & Whitburn, 1982; Paradis, 1981; Whitburn et al., 1978).

Early proponents of the plain style felt that much of the learned writing of their day was, at best, truth excessively burdened by rhetorical flourishes or, at worst, falsehood masquerading as truth in superficially fine rhetorical garments. Sprat (1958, p. 113) was of the latter opinion when he referred to rhetoric as "this vicious abundance of *Phrase,* this trick of *Metaphors,* this volubility of *Tongue,* which makes so great a noise in the World." Given this historical, almost definitional, aversion to rhetoric, it is no surprise that many technical writers today are wary of rhetoric. They, like Sprat, suspect that rhetoric is used to promote opinion, not to communicate fact.

Objections to rhetoric from Sprat's time to today are not simply expressions of stylistic preference. The controversy is one aspect of a larger philosophical debate about the nature of language itself. The theory of the plain style in technical writing assumes a univocal ideal for language: each word has a single meaning that in ideal writing would be understood by every reader without variation in denotation or even allusive nuance of connotation (Dobrin, 1983). This is the ideal world of mathematical symbology and apodictic proof. When Sprat made "Mathematical plainness" the scientific writer's goal, he "laid the groundwork for the marriage of mathematics and technical language" (Paradis, 1981, p. E-83).

The lasting appeal of Sprat's implicit analogy may be seen in Britton's (1965, pp. 114–115) much-admired definition of technical writing: "Technical and scientific writing can be likened to a bugle call . . . [its] primary . . . characteristic . . . lies in the effort of the author to convey one meaning and only one meaning in what he says." Similarly, the univocal ideal for language is the implicit premise behind the claims in many technical writing textbooks that technical writing is somehow more "objective" than other forms of writing (see Lannon, 1982; Mills & Walter, 1978; Weisman, 1980).

Acting on this premise, technical writing theorists and practitioners have excluded persuasive rhetoric from their list of appropriate stylistic concerns. Univocal writing has no need to persuade because readers cannot think otherwise than what the univocal text singularly states. In this traditional view, persuasion in technical writing can come only from the skillful application of expository strategies that seek univocal clarity by attempting to create "windowpane" texts—perfect transmitters of information—that in no way color the facts they communicate (see Miller, 1979).

But texts do color the facts they communicate, and the profession's concern with audience analysis demonstrates that technical writers recognize this. As described by Pearsall (1969), audience analysis is a method of classifying the readers of a document as technicians or executives, laypersons or experts according to the uses each reader would have for the document. After analyzing the audience, the technical writer may adjust the diction or even rearrange the order of presentation in the document so as not to insult or confuse the audience. The purpose is to give readers immediate access to the facts they seek by leveling their affective responses to the text and effacing the distracting presence of the writer. Any concern for the reader's response is potentially a rhetorical, even a persuasive, concern. But the writer-effacing character of such an audience analysis dem-

onstrates that persuasive stategies calculated to influence the reader's response to anything other than the factual content of a document have been
far from the mainstream of technical writing theory and practice.

In 1978, an entire issue of *Technical Communication* was devoted to
the role of rhetoric in technical writing. The title of the keynote article,
"Rhetoric and Technical Writing: Black Magic or Science?" (Corey, 1978,
p. 2) reflects the profession's generally suspicious attitude toward rhetoric
(see also Girill, 1985). In the one article to discuss persuasive rhetoric explicitly, Sachs (1978, p. 14) calls for a new method of audience analysis
that would use "our knowledge of psychology, of behavior modification,
of transactional analysis, and of communications" to create persuasive
rhetorical strategies whose theoretical basis would, Sachs suggests, represent an advance upon Aristotle's. In his brief article, Sachs does not suggest how these modern approaches to analyzing human behavior might be
translated into rhetorical strategies appropriate for technical writing. Sachs
is no advocate of persuasive rhetoric in technical writing. He warns that
the rhetorically "skillfull communicator" may be a "technical con-man . . .
as slick as the used-car salesman who knows how to doctor a transmission
with bananas and sawdust."

Returning to Rhetoric

In the years since 1978 academic interest in technical writing has focused
increasingly on rhetoric. First in a presentation at the 1978 Conference on
College Composition and Communication and later in print, Hairston (1982,
p. 77) imported the term "paradigm shift" from Thomas Kuhn's *The
Structure of Scientific Revolutions* (1962) to describe the "replacement of
one conceptual model by another" that has occurred in composition research over roughly the last 30 years. The traditional paradigm in composition research is based on analysis of the finished documents that writers produce. This approach is being replaced by research and pedagogy
based on analysis of writers' practices as they write. Adherents to the new
paradigm perceive writing as a process. The emerging paradigm is "rhetorically based; audience, purpose, and occasion figure prominently"
(Hairston, 1982, p. 86).

Hairston's use of Kuhn to describe the growing recognition of the importance of rhetoric was not new. The rhetorical implications of Kuhn's
concept of the paradigm shift are crucial to the arguments offered by Halloran (1978) and Miller (1979) in their two articles, written independently
but published virtually at the same time, on the rhetorical nature of technical writing. Halloran and Miller accept the basic assumption of definitions of technical writing that look back to Sprat: the scientific content of

technical writing defines the genre. Then, extending the implications of Kuhn's concept of the paradigm shift, they critique the objectivity of science itself, asserting that establishing scientific truth is in part a rhetorical process that depends upon consensual agreements reached between the authors and the readers of scientific papers.

This argument is an ironic revision of Sprat's original insight. Technical writing *is* like science in its method of truth-claiming. But neither technical writing nor science is wholly objective in its communication of truth. Both use rhetoric, even persuasive rhetoric, to convince audiences of the validity of their claims. "The test of a given scientific schema," writes Halloran (1978, p. 80), "is as much the degree to which it wins the agreement of other scientists as the degree to which it coincides with observed physical reality." And Miller (1979, p. 616) writes, "Good technical writing becomes, rather than the revelation of absolute reality, a persuasive version of experience." These assertions mark the moment of the paradigm shift in the study of technical writing. Since Halloran and Miller, most analyses of technical writing proceed from this rhetorical assumption about the nature of technical writing (see Anderson, Brockman, & Miller, 1983; Bazerman, 1983; Dobrin, 1982, 1983, 1985; Zappen, 1983).

Consider the explicitly rhetorical definition of technical writing given by Dobrin (1982, p. E-40): technical writers "mediate between groups who don't have a shared understanding" of a document's subject. Dobrin's concept of technical writing as the textual mediation by authors to readers seeking understanding is nothing like Britton's (1965, pp. 115, 114) ideal "bugle call" writing with "one meaning and only one meaning." Dobrin's definition suggests that meaning in technical writing is similar to the equivalent, rather than exact, meaning readers derive from translations. Such meaning is relative and contextual; it is not, as Britton would ideally have it, one to one and exact.

In this context, the study of persuasive rhetoric in technical writing discovers textual structures that foster the mediation of meaning between authors and readers. To be truly effective, such a "new rhetoric" of persuasion could not, as Sachs (1978) feared, be a flim-flam scheme designed to hoodwink innocent readers. Instead, as Burke (1969, p. 43) argues, the new persuasive rhetoric is "rooted in an essential function of language itself . . . the use of language as a symbolic means of inducing cooperation in beings that by nature respond to symbols." Similarly, Booth (1974, p. 137) writes that "the supreme purpose of persuasion in this view could not be to talk someone else into a preconceived view; rather, it must be to engage in mutual inquiry or exploration." The most effective and persuasive technical writing encourages readers to approach functional documents in this spirit of mutual inquiry and exploration. Only then can authors communicate to readers the seemingly objective truth that comes when the participants in a frank and honest discussion are persuaded to accept a consensus opinion.

Mediation as Persuasion: Theory for a New Rhetoric

Mediation is the key concept for understanding how rhetorical strategies that seek to persuade readers by engaging them in mutual inquiry are fundamental to technical writing. Such persuasive rhetoric cannot be achieved through alterations of style alone. Manipulating a text for the sole purpose of persuading its audience is false rhetoric. Plato attacked this practice in the *Gorgias,* and Aristotle sought to correct it in the *Rhetoric* (see Corbett, 1971, pp. 31–39, 595–599; Kinneavy, 1980, pp. 5–8, 211–228).

Since Sprat's time, technical writing has opposed false rhetoric with plainness and objectivity. But a plain style and an objective tone can promote false rhetoric as surely as the most ornate rhetorical amplification. When writers mechanically adjust style to suit readability formulas or use passive voice constructions such as "it was observed that" and "it was concluded that" to promote an air of objectivity, they attempt to persuade readers through superficial manipulations of style that a document is objective and therefore true. Used in this manner, readability formulas and passive voice constructions are falsely persuasive rhetorical strategies. Their use is hypocritical and, as research has shown, ineffectual (see Macdonald, 1983; Redish & Selzer, 1985; Selzer, 1983; Sides, 1982).

In contrast, persuasion deriving from mediation uses audience analysis to promote understanding, both intellectual and affectual, and then persuades through the mediation that occurs in the act of understanding. Aristotle's analysis of rhetoric assumes a similar progression from understanding to persuasion; this progression alone is not what makes the "new rhetoric" new (see Lunsford, 1979, p. 150). The new rhetoric explicitly states that persuasion comes from the mutual understanding of authors and readers; it values rhetorical strategies that communicate understanding: the more effective the communication, the more persuasive the rhetoric. Aristotle limited the scope of rhetoric to "such things as come, more or less, within the general ken of all men and belong to no definite science. . . . [2154] the approximately true" (Barnes, 1984, p. 2152, 2154; see also Kinneavy, 1980, pp. 218–228). Rhetorical inquiry today considers "the whole field of the uses of language" (Kinneavy, 1980, p. 3), acknowledging the problematical relationship between language and truth. Emphasizing the communication process expands the field of rhetorical inquiry to include every discipline that seeks to discover and communicate knowledge. This expansion of rhetorical concern is what makes the new rhetoric new (see Burke, 1969, pp. 3–46; Kinneavy, 1980, pp. 1–5, 48–72).

To be rhetorically effective, a document must mediate between the logic and desires of a writer and the presumed logic and desires of the writer's audience. A document is rhetorically effective when its readers are persuaded that they share a community of logical and affectual understanding with the writer; the document communicates through its ability to promote the "community" of the author and the audience. So, too, does a document persuade. Once readers understand the writer's point of view and believe that the writer understands theirs, they may accept or reject the document's message on the basis of their own judgment. Given the new rhetorical understanding of the nature of language itself, that judgment is as objective as any judgment can be. This is the structure of true rhetorical persuasion. It is as true for technical writing as it is for any other form of writing.

Persuasion and Truth

The new rhetoric of persuasion challenges our understanding of what constitutes truth. The fullest statement of the role that rhetoric plays in the establishment of truth may be found in the work of the Belgian philosophers Chaim Perelman and L. Olbrechts-Tyteca (for discussion see Dearin, 1969, 1982). Perelman and Olbrechts-Tyteca (1971, p. 1), attempting to rehabilitate rhetoric, attack the Cartesian, positivist tradition in philosophy for having made "the self-evident the mark of reason, and considered rational only those demonstrations which, starting from clear and distinct ideas, extend, by means of apodictic proofs, the self-evidence of the axioms to the derived theorems." This attack on the Cartesian conception of what constitutes a valid demonstration of truth is also an attack on the objective truth-claim of the sciences.

Writing alone in an earlier article, Perelman (1963, p. 94) asserts that science claims to be

> composed of self-evident truths fixed *ne varietur*, whatever the further development of knowledge: this assumes that the language in which these truths are enunciated, and the ideas that serve to express them, will be subjected to no future reversal in consequence of the progress that science might make.

Perelman anticipates the argument for the rhetorical nature of scientific truth in Kuhn's *The Structure of Scientific Revolutions* (1962) when he suggests that scientific progress does not follow Cartesian, positivist principles for demonstrating the truth of propositions:

> [I]f we assume that the sciences develop on the basis of opinions previously accepted—and replaced by others either when difficulty results from some

contradiction or in order to allow for new elements of knowledge being integrated in the theory—then the understanding of scientific methodology requires us to be concerned not with building the scientific edifice on the foundation of self-evident truths, but with indicating why and how certain accepted opinions come to be no longer regarded as the most probable and the most suitable to express our beliefs, and are replaced by others. The history of the evolution of scientific ideas would be highly revealing in this regard. (Perelman, 1963, p. 94)

What would be historically detailed in Kuhn's later analysis, and later remarked upon by Hairston (1982), Halloran (1978), and Miller (1979), is the central role that persuasive rhetoric has played in the development of what we accept as scientific truth.

Truth and Technical Writing

Perelman's argument for the validity of rhetorical persuasion in philosophical discussions bears upon our argument that persuasive rhetoric is appropriate in technical writing. In both cases, the crux of the issue is how one understands the process whereby meaning is communicated in language. Does one accept the concept of a univocal ideal or opt instead for a pragmatic, rhetorical approach? Perelman (1982, p. 296) illustrates the fundamental divergence that occurs at this point by comparing the positions taken by Plato and Aristotle:

> Whereas for Plato any idea can be the object of intuition, which reduces philosophical rhetoric to the subordinate role of communication of a truth acquired independently, . . . for Aristotle only theoretical principles can be the object of intuition and of scientific knowledge, extending thereby enormously the field of rhetoric and dialectic, [they are] alone capable of justifying our decisions and our choices. The realm of rhetoric will be extended still more if one stresses the fact that intuitions cannot be expressed without the language of a cultural community.

We may lift Perelman's argument out of its context of philosophical debate and into our discussion of truth in technical writing by substituting "objective data," the common content of technical writing, for Perelman's "object of intuition," and "technical writing" for his "philosophical rhetoric." Rereading the passage in paraphrase, Plato would belong in Sprat's and Britton's camp, calling for technical writing that exhibits the "Mathematical plainness" of a "bugle call" and it would be limited "to the subordinate role of communication of a truth acquired independently."

If technical writing were solely the communication of independently acquired truth, technical writers would be replaced by computers, relaying with unerring accuracy bits and bytes of data from one machine to another. But the need for technical writers persists. After data are compiled, analyzed, and reported, technical writers must create documents that communicate those data in a context so that readers can make informed decisions about the data's use. Technical content may be objectively represented with more or less "bugle call" accuracy in an equation, in a plain table of experimental results, or even in a simple graph. But the text of a functional document must translate objective data into a subjective context. In a document's reader-oriented context of use, rhetorical strategies will likely help justify the validity of the decisions readers will make about the data's use. In the writing of the document, persuasive rhetoric may ethically urge the truth-claims of the various choices that must be considered before those decisions are made.

The technical writer's professional expertise is to take the objective truth of technical data and translate it into language that persuades readers of the data's truth in the context in which the data would be used. In any functional document, technical writers translate the objective data of the sciences into the rhetorical language of documents that must convince readers how the objective data they communicate should be applied. Technical writers belong in the class of those for whom the "habitual medium is the word"; they are, like many other professionals and scholars, "rhetoricians, though they may not know it" (Raymond, 1982, p. 780).

Recommendations

To make persuasion through mediation work, writers must invite their readers to consider the subject of a document in the spirit of "mutual inquiry or exploration" that Booth (1974, p. 137) described.

The structure of such a document's text would act as a mediating bridge, a rhetorical construct, spanning the communication gap between the writer and the reader. The "bridging" metaphor for describing the technical writer's craft is not new. The first sentence of the Society for Technical Communication's (STC) "Code for Communicators" (no date) affirms that "As a technical communicator, I am the bridge between those who create ideas and those who use them." This could as easily be the code of a hypothetical "Society of Professional Rhetoricians." However, as we have seen, many STC members prefer to equate the bridge building in their writing with the work of civil engineers rather than the textual constructions of rhetoricians. Nevertheless, by comparing two examples of a single rhetorical strategy that persuades through mediation, we can discover an effective and ethical method for making technical writing more persuasive.

Rogerian Argument and Technical Persuasion

Persuasive rhetoric based on mediation appeals to our desire to be understood and our need for communication. One such rhetorical strategy has been derived from the work of the American psychologist Carl Rogers. Rogers (1970, pp. 204–285) suggests that

> The emotionally maladjusted person, the "neurotic," is in difficulty first because communication within himself has broken down, and second because as a result of this his communication with others has been damaged. . . . We may say then that psychotherapy is [fostering] good communication. The major barrier to mutual interpersonal communication is our very natural tendency to judge, to evaluate, to approve or disapprove, the statement of the other person, or the other group. . . . Real communication occurs, and this evaluative tendency is avoided, when we listen with understanding. What does that mean? *It means to see the expressed idea and attitude from the other person's point of view, to sense how it feels to him, to achieve his frame of reference* [Rogers's emphasis].

Rogers's emphasis on communicating "the expressed idea and attitude from the other person's point of view" demands a subtle but unambiguously rhetorical audience analysis. The therapist's goal is to influence favorably the client's perception of the therapist and, thusly, to gain the client's trust.

The principles of Rogers's client-centered therapy have been adapted to rhetorical theory and named *Rogerian argument* (Young, Becker, & Pike, 1970, pp. 273–283). The persuasiveness of Rogerian argument depends on a writer's ability to communicate to the reader not only an understanding of the reader's point of view but also that the writer's own position is substantially informed by that understanding (see Flower, 1981, pp. 143–166). The writer then invites the reader to consider both the shared and the divergent aspects of their positions. Writer and reader share the common goal of achieving a fuller understanding of the question. To be successful, Rogerian argument must demonstrate constant respect for the reader's point of view. Its persuasiveness lies in its appeal to cooperation over conflict, to mediation in search of understanding, and to the need "either [to] build or [to] discover bridges . . . that will encourage trust" (Young, Becker, & Pike, 1970, p. 280).

There have been objections to using Rogerian argument as a persuasive strategy in technical writing. Stoddard (1985, p. 231) suggests that Rogerian argument

> is insufficiently complex to apply to the competitive economy within which technical writing transpires. When Northrop writes a proposal for a defense

contract, it is not seeking "social cooperation" but victory over competing firms. In the best of all possible worlds, Rogerian argument would be the best rhetoric, but in a capitalist marketplace based on individual competition, it can only be an ideal.

Stoddard's objection that Rogerian argument is "insufficiently complex" for the competitive world of technical writing depends on a narrow view of what "victory" means in the context of proposal writing. For Stoddard, using rhetoric to persuade an audience and win a proposal means controlling the audience's point of view. She associates this practice with Aristotelian rhetoric. Stoddard's purpose is to offer a revised Aristotelian rhetoric that will retain the competitiveness she associates with Aristotle but still be ethical and, therefore, suitable for technical writing. Stoddard implies that Rogerian argument is too soft, too humanistic. Its vague notion of the persuasiveness of "social cooperation" isn't realistic in the high-tech world where what matters is that you win, not how you play the game.

But what really happens when Northrop wins a proposal for a defense contract? Northrop does not vanquish the competition by coercing the government to discount the merits of competing proposals. All of the competing proposals will be read very carefully, and the winning proposal will be chosen because it fits best what the request for proposal (RFP) seeks. Organizations issue RFPs to seek out contractors capable of understanding and then addressing their particular needs. Proposal writers must construct documents that promise to deliver mutually beneficial cooperation between client and contractor—a reasonable match of need with ability to satisfy that need. Northrop wins defense contracts by writing proposals that identify and respond to the government's needs more fully than the proposals of its competitors.

Advice on proposal writing emphasizes the importance of audience analysis. A proposal writer must see the company "through the client's eyes" (Whalen, 1982, p. 19) and write a proposal that communicates to the client first that the client's needs are understood and then that the contractor can satisfy them. Successful proposals "form and shape the client's perception of what is realistic" (Whalen, 1982, p. 7), often even beyond the client's understanding of the situation as it is expressed in the RFP (see Whalen, 1985).

Expanding and shaping a client's perception of the problem are the bases of the persuasive rhetoric of Rogerian argument. As the therapist, Rogers must persuade a client to see the problem as the therapist sees it; in the same manner, the proposal writer must persuade a potential client to see the proposal situation as the proposal writer sees it. The goal in both cases is to influence the client's perception of the situation so that the client sees the therapist (or the proposal writer) as offering what the client seeks: a professional, mutually beneficial, cooperative relationship, attested by a contract that promises to accomplish whatever action the situation demands. Persuasion based on the offer of social cooperation is not just an ethical way to play the game of proposal writing. It is the name of the game. And it wins.

The Structure of Rogerian Argument

When Young, Becker, and Pike (1970, p. 275) first proposed Rogerian argument as an alternative to Aristotelian persuasive rhetoric, they suggested that it has "no conventional structure . . . because these [structural] devices tend to produce a sense of threat, precisely what the writer [who uses Rogerian argument] seeks to overcome." Hairston (1976) has offered a five-part structure for adapting the rhetorical principles of Rogerian argument to the writer's task of composing an argumentative document. Because our concern is with persuasion in general, not with the formal procedures of argumentation, we can reduce Hairston's five-part structure for Rogerian argument to a three-step progression.

Step 1: State the facts of the situation in as neutral a context as possible. Avoid evaluating the situation. Use third-person, passive voice verb constructions to focus the reader's attention impersonally on the situation itself, on the action(s), not on the actor(s) involved.

Step 2: Demonstrate understanding of the reader's point of view by discussing the document's subject in the context of the reader's interests. Use active voice, second- or third-person verb constructions that portray the audience as the actor. Focus attention on the reader's action and on your perception of the reader's role. Then, shifting from second-person to first-person plural verb constructions, unite your action(s) and perception(s) with the reader's, establishing a common ground of mutual understanding and concern.

Step 3: Finally, state your aims or goals or analysis of the situation boldly and straightforwardly in the context of mutual understanding and concern established in steps 1 and 2. Make value judgments and ask for what you want. Use first-person singular and plural verb constructions.

The first step, an unemotional statement of the facts of the situation, appeals to the technical reader's immediate desire for data. By not evaluating the data, the writer invites the reader to evaluate the situation as a neutral, objective observer. The persuasive strategy is to encourage the reader to regard the writer as a similarly objective observer.

The second step is the most difficult. Having established credibility and objectivity, the writer must establish a subjective understanding of the reader's point of view. This is the bridge-building step. The writer must perform a sophisticated audience analysis to scrutinize the reader's needs and the purpose of the document. The writer then discusses the document's subject as much as possible within the context of the reader's interest in and probable use of the information communicated, demonstrating that the reader's interest in the subject is similar to the writer's interest. Finally, the writer must believe that the context of trust and understanding is sufficiently strong to allow a direct statement of what the writer wants.

This three-step, persuasive structure derived from Rogerian argument fits well with the most fundamental rhetorical structure of exposition: a document should have a clear introduction that announces the subject and suggests its significance; a complete middle section that develops the significance of the subject; and a satisfying conclusion that explicitly states the significance of the subject in the fully developed context of the document. In the Rogerian progression, step 1 is well suited for an introduction that sets the context for a document; step 2 may be satisfied in the expository course of the body, or development, of a document; and step 3 lends itself naturally to the expression of conclusions and recommendations.

The three steps of Rogerian persuasive rhetoric may also be understood as an adaptation of Aristotle's three modes of rhetorical persuasion: logos, pathos, and ethos.

Aristotle's Categories	Technical Writing Context
Logos: Nature of the message	Quality of the data
Pathos: Emotion of the audience	Potential benefit to the audience from understanding and applying the data
Ethos: Nature of the source	Integrity of the author or organization: technical competency and objectivity

Figure 3.1 Aristotle's Categories and Technical Writing

Figure 3.1 relates Aristotle's modes of persuasion to technical writing, listing them in the order of their importance as persuasive appeals according to the traditional approach to persuasion in technical writing. Ethos, the character of the writer, is listed last. It could almost be removed from the figure entirely. The technical writer's ethos, or integrity, is traditionally seen as an implicit function of the document's appeal to logos, the quality of the data as they are presented by an impersonal exposition that reflects "the ideal of scientific objectivity" (Stoddard, 1985, p. 229).

Rather than redefine Aristotelian rhetoric, Rogerian persuasion redefines the relative importance to technical writing of the three Aristotelian modes of persuasion. In the Rogerian scheme, ethos is returned to its classical status as the most important mode. Aristotle observes that "we believe good men more fully and readily than others. . . . character may almost be called the most effective means of persuasion" (Barnes, 1984, p. 2155). The technical writer employing Rogerian persuasion wants readers to understand appeals to pathos (assessments of the readers' use of the data derived from audience analysis) and appeals to logos (the quality of the data itself) in the established context of the writer's ethos, the writer's competent, scientifically objective character.

The three steps of Rogerian persuasion may be applied appropriately in several contexts. Let us compare two distinctly different documents: a very short poem, "This Is Just to Say," by the modern American poet William Carlos Williams, and a memo that proposes to an oil company a research and development project involving the mining of mineral-rich sediments from the Red Sea.

An Unlikely Example

Williams's poem conveniently reflects the structure of Rogerian persuasion in three succinct paragraphs of four lines each. Its three paragraphs can be juxtaposed with the three parts of Rogerian persuasive rhetoric.

This Is Just to Say

I have eaten the plums that were in the icebox	*THREE*	State the facts of the situation in as neutral a context as possible. Avoid evaluating the situation.
and which you were probably saving for breakfast	*PART*	Demonstrate understanding of the reader's point of view. Use second-person verb constructions.
Forgive me they were delicious so sweet and so cold	*STRUCTURE*	State what you want in the context of your mutual understanding of the situation.

The speaker's persuasive goal is clear enough. The poem is a confession; the speaker asks for forgiveness. The dramatic situation is also apparent. The reader is eavesdropping, reading a brief note, perhaps a domestic memo from one spouse to another. The note itself might be found taped to the door of the refrigerator they share. The language is plain, and the message is straightforward.

For all its charming simplicity, the persuasive strategy behind the speaker's confession and request for forgiveness is as subtle as the potential conflict the poem seeks to avert. Suppose the speaker had begun with the first line of the last paragraph, *Forgive me?* Or worse, started from the top extolling the *delicious / so sweet / and so cold* joys of eating the plums? Either choice would have made an immediate evaluation of the situation. To write *Forgive me* in the first line would emphasize the speaker's guilt and cast the spouse in a superior role. To go on about how delicious the plums were would have the opposite effect. The speaker would seem unashamed, rubbing in the spouse's loss of the *delicious* fruit. No part of the poem can be moved without reducing the effectiveness of the speaker's persuasive appeal.

The speaker's simple statement of the facts of the situation in the first verse paragraph defines the problem without making any value judgments that might suggest how either the speaker or the spouse should respond. The effect is to open the subject in a neutral manner, to portray the speaker as an ethical, honest person with nothing to hide. Saying *I have eaten the plums* does not admit to wrongdoing or guilt; the speaker avoids admitting that the plums should not have been eaten by introducing the guilty action in as neutral a context as possible. The speaker's purpose is to persuade the spouse to withhold judgment, to read on and hear more before responding.

In the second-verse paragraph, the speaker acknowledges guilt, but the context demonstrates that the speaker understands the spouse's point of view. The spouse was *probably / saving* the plums because the spouse also must have recognized their value. The speaker establishes a common ground of mutual understanding—they both enjoy plums—and the implication is that the spouse would have acted exactly as the speaker did. The persuasive strategy is to portray the speaker and the spouse as equally sensitive, understanding people. The speaker demonstrates ethical character by a full, unforced confession. When the speaker steps forward in the final paragraph and asks the spouse directly, *Forgive me,* forgiveness seems likely.

The spouse is persuaded by the structure of the poem, the rhetorical order of the presentation, to forgive the speaker. The rhetoric mediates the speaker's desire for the plums with the similar desire of the spouse by establishing the common ground of their shared experience of the pleasures of eating plums. The literary critic Charles Altieri (1979, p. 497) suggests that "This Is Just to Say" presents an ideal rhetoric that promotes understanding and the resolution of conflict in "an arena where [men and women] can discover the joys of making and testing truths as a necessary communal enterprise." In this statement we should hear echoes of Booth's understanding of persuasion that derives from "mutual inquiry," echoes of Perelman's and Kuhn's understanding of the rhetorical nature of scientific truth, and echoes of the persuasion based on mediation that may be derived from Rogerian argument.

Persuasion in Practice

The memo below was written by a mining engineering student as a cover memo to a proposal abstract. The writer proposes that he research and then report on the feasibility of a speculative, undersea mining project. Probably influenced by the rhetoric of many advertisements, the writer uses a negative appeal in his attempt to be persuasive. We are all familiar with persuasion based on a negative appeal. We might lose our job if we recommend the wrong computer to our employer suggests one advertisement; another implies that we will have no social life if we use the wrong toothpaste.

Negative appeals approach the persuasive process in a manner exactly opposite to Rogerian argument. The writer of a negative appeal immediately evaluates the situation and suggests how readers should respond to it, threatening them if they do not accede to the writer's point of view. Much of our modern suspicion of persuasive rhetoric probably derives from our rightly skeptical responses to advertising that attempts to influence us by playing on our fears. Unfortunately, when we try to write persuasively, we often employ the rhetoric with which we are most familiar, as the writer of the example memo does below.

> Oilco cannot afford to ignore the accompanying abstract. It should prove to be of great interest to our company because of its application to both mineral processing and oil production technology. Indeed, it is probable that the proposed program could only be undertaken by a company of our size and diversity. Thus the competition is minimized. The program involves research and development of the largest Red Sea geothermal deposit, the Atlantis II Deep. This deposit consists of unconsolidated sediments in several mineralogical facies of varying metallic content. These sediments would be recovered by fluidization of the facies. The material would then be pumped to a surface platform. Oilco now has the great opportunity to develop and patent fluidization technology. This will enable us to become the sole purveyor of this technology which must be used to profitably extract ores from various marine deposits. Our position astride the oil and mineral extraction fields puts us in the best position to exploit such marine deposits. I feel that this abstract deserves the most serious consideration by you and your staff.

The writer takes an advertiser's sweepstakes approach. Oilco *cannot afford to ignore* this great idea; they will be sorry if they do not take a chance on it. That may be true. But sweepstakes advertisements are mailed to millions of people, and thousands will enter. Most will throw the mailing away. Because this memo will be evaluated initially by a single person, daring that reader to ignore the proposal is likely to result in the dare being accepted and the document being ignored. The writer risks losing the audience's good will in the first sentence. The content of the memo is actually quite strong. The idea is new and interesting, and there is ample justification for pursuing it. The problem is the rhetoric.

Without changing the content of the memo, we can rearrange the order of the presentation of the information to fit the three-part, rhetorical structure of Rogerian persuasion. Remembering the three paragraphs of Williams's poem, let's revise the memo's one paragraph into three paragraphs, each presenting one part of the Rogerian strategy. The first paragraph should contain the nonevaluative, expository sentences that detail what the project is about and how it would work. This material represents the facts of the situation. In the second paragraph we will fit Oilco into the picture by using those sentences that discuss how Oilco's dual corporate interests in mining and oil production make them the best organization to pursue this project. The explicitly evaluative sentences remain for the third paragraph. In the revised memo we need to add one sentence that asks directly for permission to begin research. With a few minor changes in the sentence structure, some equally minor stylistic editing, and no substantive revisions of the original memo's factual content, the revised memo reads competently, professionally, and persuasively.

> The accompanying abstract proposes research and development of the largest geothermal mineral deposit in the Red Sea, the Atlantis II Deep. This deposit consists of unconsolidated sediments in several mineralogical facies of varying metallic content. These sediments would be recovered by fluidization of the facies. The material would then be pumped to a surface platform.
>
> Oilco's position astride both the oil and the mineral extraction fields puts us in a good position to develop and apply the necessary mineral processing and oil production technology. Because the proposed program probably can be undertaken only by a company of Oilco's size and diversity, our competition would be minimal. We could possibly become the sole purveyor of a new extraction technology and the world leader in marine mining.
>
> Oilco cannot afford to ignore the accompanying abstract. I feel that it deserves the most serious consideration by you and your staff. May I begin researching the project as soon as I have your approval?

Paragraph 1 states the facts of the situation with no implicit evaluation of the proposal's merit. The writer focuses on the subject of the memo without alluding to his or his reader's interests in it. The writer tells what will be done, where it will be done, and how it will be done; he does not say who will do these things. The writer avoids personal pronouns, casting the *abstract,* the *deposit,* the *sediments,* and the *material* as the grammatical subjects of the sentences. These impersonal sentence constructions emphasize the actions proposed, not the actors who will carry them out. The writer's persuasive strategy is to allow the reader to evaluate the situation. The Rogerian tactic is to avoid any evaluative, presumptive posture that might put the reader on guard. The writer portrays himself as an objective, disinterested source of information. The Aristotelian rhetorical appeal is to establish the writer's professional competency (ethos) and to portray the writer as a trustworthy colleague who understands the objective ideal (logos) that governs the practice of professional technical communication.

In paragraph 2, the writer establishes that he understands the audience's point of view: Oilco's corporate concerns and capabilities match what is needed to carry out this project successfully. Oilco is the subject of the first sentence in the paragraph, and the writer uses the first-person plural pronoun *us* in the object position to mark grammatically the identity of his interests with Oilco's interests. Oilco's potential for profit, its stake in the matter, is implied as the writer continues the grammatical and rhetorical shift from impersonal, third-person singular verb constructions to an inclusive, first-person plural predicate: . . . *our competition would be minimal. We could become the sole purveyor of a new extraction technology.* The writer takes the second step in the Rogerian persuasive strategy by emphasizing that he and the readers have a mutual interest in the proposed project. They both will gain by pursuing their shared goal. The Aristotelian persuasive tactic in this paragraph is an emotional appeal (pathos) to Oilco's desire for profit.

The most threatening sentence from the first draft of the memo reappears at the beginning of the third paragraph: *Oilco cannot afford to ignore the accompanying abstract.* Although the sentence is identical in both versions of the memo, the rhetoric of the revised version changes its formerly threatening posture to an expression of shared urgency. As the first sentence in the original memo, the phrase *cannot afford to ignore* was a vague threat because the readers had no idea what they were being dared not to ignore. In the revised memo, the readers know what the writer is proposing and have probably made some objective judgment of the probable merit of both the proposer and the proposal. The first paragraph of the revised memo establishes the trustworthy character of the writer, and the second paragraph links the writer's goal with the company's goal. The readers can decide whether they will ignore the proposal. What had been a threat becomes an expression of the writer's strong support for a project that he expects his readers to understand. Finally, the writer concludes by stating in a straightforward manner what he wants from the reader: *May I begin researching. . . .*

Conclusion: A Larger Context

Adapting Rogerian argument into a persuasive rhetoric appropriate for technical writing is one example of the practical benefits that technical writers may expect from rhetorical studies of their craft. The average person thinks of professional writers as creative writers—poets, essayists, and novelists—writers who create the worlds that their texts communicate to a general audience. Although technical writers also create in this manner to some extent, they are more appropriately thought of as professional rhetoricians, as writers who take data from a variety of information sources and construct from them documents that effectively address a specific audience. Technical writers know that much technical writing is technical rewriting, the revision of draft after draft as new documents are created by cutting and pasting, reordering and rewriting the information presented in prior documents. Rhetorical strategies are the basic tools for such textual construction work. Unfortunately, rhetoric, persuasive or otherwise, is still regarded with suspicion by many professional technical writers.

Writing in *Technical Communication,* Lay (1984, p. 12) encouraged the use of the enthymeme (a syllogism with an implicit major premise) as a rhetorical device that could help technical writers "clarify our logic and state clearly our main and supporting assertions, . . . encourag[ing] the reader to participate in our argument by supplying the hidden assumptions in [the] enthymemes." Encouraging readers to participate in the development of a document's argument is one means of linking persuasion with exposition in a manner similar to the method of Rogerian argument.

In a later issue, Sachs (1984, p. 8) attacked Lay's suggestion on the grounds that "writers may leave the premise out because the purpose of the enthymeme is to trick the reader." Sachs then jumped from his conditional statement about the unethical practices of some writers to dismiss illogically any use of the enthymeme: "the enthymeme conceals false premises and its use should be considered unethical. That hardly makes it suitable for technical communications." Aristotle long ago pointed out the fallacy of blanket dismissals of rhetoric: "What makes a man a sophist [a false rhetorician] is not his abilities but his choices" (Barnes, 1984, p. 2155). Sachs (1984, p. 12) coyly challenges the reader to recognize his purposefully faulty logic, "Did you . . . supply the premises, or were you fooled?" offering his own argument as an example of why enthymemes should not be used in technical communication. But Sachs' antirhetorical prejudice is clear.

The problems that come of dismissing rhetoric from technical writing extend beyond limiting the tools that technical writers may employ in their trade. There can be minimal professional pride or development if technical writers define their practice as no more than trying to stay out of the way of the objective truth communicated in the documents they write. The technical writer's role is much more active and far more influential than that. A professional occupation should derive the principles of its practice from a "generalized cultural tradition" (Parsons, 1967, p. 536). Technical writing belongs to the cultural tradition of the humanities; the principles of its practice have developed as one branch of the history of rhetorical concern for accurate and effective writing.

Acknowledging the importance of rhetorical study to technical writing would bring to the center of attention persuasive tactics for presenting information honestly and objectively in a compelling manner. It would also focus more attention on the ethical problems that technical writers face when, through the documents they write, they act as anonymous spokespersons for a variety of businesses, research laboratories, and the government. Technical writers should play more than a peripheral role in addressing the important challenges to communication and understanding posed by our increasingly complex technology. We need a new rhetoric that will speak to our highest ethical qualities with enough persuasive power to promote the deepest understanding and the wisest use of the technological devices we continue to fashion from our growing mastery of science.

References

The research for this essay was begun in the summer of 1984 when I attended the National Endowment for the Humanities Summer Seminar, "Literary Theory and the Romantic Self," directed by Professors Morris Eaves and Michael Fischer. I thank Professors Eaves and Fischer for their encouragement, and I thank the National Endowment for its support. Also, a portion of this essay was published in a different form as, "Building Better Bridges: Persuasive Structure and Technical Writing," in the *Proceedings* of the 34th International Technical Communication Conference (May 1987) sponsored by the Society for Technical Communication.

Altieri, C. (1979). Presence and reference in a literary text: The example of Williams's 'This is just to say,' *Critical Inquiry, 5,* 489–510.

Anderson, P. V., Brockman, R. J., & Miller, C. R. (1983). Research in technical and scientific communication. In P. V. Anderson, R. J. Brockman, & C. R. Miller (Eds.), *Essays in technical and scientific communication: Research, theory, and practice* (pp. 7–14). Farmingdale, NY: Baywood.

Barnes, J. (Ed.). (1984). *The complete works of Aristotle: The revised Oxford edition,* Vol. 2 (W. R. Roberts, Trans.). Princeton, NJ: Princeton University Press.

Bazerman, C. (1983). Scientific writing as a social act: A review of the literature of the sociology of science. In P. V. Anderson, R. J. Brockman, & C. R. Miller (Eds.), *Essays in technical and scientific communication: Research, theory, and practice* (pp. 156–184). Farmingdale, NY: Baywood.

Booth, W. C. (1974). *Modern dogma and the rhetoric of assent.* Chicago: University of Chicago Press.

Britton, W. E. (1965). What is technical writing? *College Composition and Communication, 16,* 113–116.

Burke, K. (1969). *A rhetoric of motives* (pp. 3–46). Berkeley: University of California Press.

Corbett, E. P. J. (1971). *Classical rhetoric for the modern student* (2nd ed.). New York: Oxford University Press.

Corey, R. L. (1978). Rhetoric and technical writing: Black magic or science? *Technical Communication, 25,* 2–6.

Dearin, R. D. (1969). The philosophical basis of Chaim Perelman's theory of rhetoric. *Quarterly Journal of Speech, 50,* 213–224.

Dearin, R. D. (1982). Perelman's concept of "quasi-logical argument": A critical elaboration. In J. R. Cox & C. A. Willard (Eds.), *Advances in argumentation theory and research* (pp. 78–94). Carbondale: Southern Illinois University Press.

Dobrin, D. N. (1982). What's wrong with the mathematical theory of communication? In *Proceedings of the 29th International Technical Communication Conference* (pp. E37–40). Washington, DC: Society for Technical Communication.

Dobrin, D. N. (1983). What's technical about technical writing? In P. V. Anderson, R. J. Brockman, & C. R. Miller (Eds.), *Essays in technical and scientific communication: Research, theory, and practice* (pp. 227–250). Farmingdale, NY: Baywood.

Dobrin, D. N. (1985). Is technical writing particularly objective? *College English, 47,* 237–251.

Flower, L. (1981). *Problem-solving strategies for writing.* New York: Harcourt, Brace & World.

Girill, T. R. (1985). Among the professions: Technical communication and rhetoric. *Technical Communication, 32,* 44.

Hairston, M. (1976). Carl Rogers's alternative to traditional rhetoric. *College Composition and Communication, 27,* 373–377.

Hairston, M. (1982). The winds of change: Thomas Kuhn and the revolution in the teaching of writing. *College Composition and Communication, 33,* 76–88.

Halloran, M. S. (1978). Technical writing and the rhetoric of science. *Journal of Technical Writing and Communication, 8,* 77–88.

Halloran, M. S., & Whitburn, M. D. (1982). Ciceronian rhetoric and the rise of science: The plain style reconsidered. In J. J. Murphy (Ed.), *The rhetorical tradition and modern writing* (pp. 58–72). New York: Modern Language Association.

Holtz, H. (1979). *Government contracts: Proposalmanship and winning strategies.* New York: Plenum.

Jones, R. F. (1930). Science and English prose style in the third quarter of the seventeenth century. *PMLA, 45,* 977–1009.

Kinneavy, J. L. (1980). *A theory of discourse.* New York: Norton.

Kuhn, T. S. (1962). *The structure of scientific revolutions*. Chicago: University of Chicago Press.

Lannon, J. M. (1982). *Technical writing* (2nd ed.). Boston: Little, Brown.

Lay, M. M. (1984). The ethymeme: A persuasive strategy in technical writing. *Technical Communication, 31,* 12–15.

Lunsford, A. A. (1979). Aristotelian vs. Rogerian argument: A reassessment. *College Composition and Communication, 30,* 146–151.

Macdonald, N. H. (1983). The UNIX Writer's Workbench software: Rationale and design. *The Bell System Technical Journal, 62,* 1891–1908.

Miller, C. R. (1979). A humanistic rationale for teaching technical writing. *College English, 40,* 610–617.

Mills, G. H., & Walter, J. A. (1978). *Technical writing* (4th ed.). New York: Holt, Rinehart & Winston.

Paradis, J. G. (1981). The royal society, Henry Oldenburg, and some origins of the modern technical paper. In *Proceedings of the 28th International Technical Communication Conference* (pp. E-82–86). Washington, DC: Society for Technical Communication.

Parsons, T. (1967). Profession. In David Sills (Ed.), *Encyclopedia of sociology,* Vol. 12 (pp. 536–547). New York: Macmillan/Free Press.

Pearsall, T. E. (1969). *Audience analysis for technical writing*. Beverly Hills, CA: Glencoe.

Perelman, C. (1963). The role of decision in the theory of knowledge. In J. Petrie (Trans.), *The idea of justice and the problem of argument* (pp. 88–97). New York: The Humanities Press.

Perelman, C. (1982). Philosophy and rhetoric, J. F. Merryman (Trans.). In J. R. Cox & C. A. Willard (Eds.), *Advances in argumentation theory and research* (pp. 287–297). Carbondale: Southern Illinois University Press.

Perelman, C., & Olbrechts-Tyteca, L. (1971). J. Wilkinson & P. Weaver (Eds. and Trans.), *The new rhetoric: A treatise on argumentation*. Notre Dame, IN: University of Notre Dame Press. (Original work published 1958.)

Raymond, J. C. (1982). Rhetoric: The methodology of the humanities. *College English, 44,* 778–783.

Redish, J. C., & Selzer, J. (1985). The place of readability formulas in technical communication. *Technical Communication, 32,* 46–52.

Rogers, C. (1970). Communication: Its blocking and its facilitation. In R. E. Young, A. L. Becker, & K. L. Pike (Eds.), *Rhetoric: Discovery and change* (pp. 284–289). New York: Harcourt, Brace & World.

Sachs, H. (1978). Rhetoric, persuasion, and the technical communicator. *Technical Communication, 25,* 14–15.

Sachs, H. (1984). The ethics of ethymemes [Letter to the editor]. *Technical Communication, 31,* 8, 12.

Selzer, J. (1983). What constitutes a "readable" technical style? In P. V. Anderson, R. J. Brockman, & C. R. Miller (Eds.), *Essays in technical and scientific communication: Research, theory, and practice* (pp. 71–89). Farmingdale, NY: Baywood.

Sides, C. H. (1982). Reassessing readability formulas. In *Proceedings of the 29th International Technical Communication Conference* (pp. E-106–109). Washington, DC: Society for Technical Communication.

Society for Technical Communication (no date). Code for communicators. Washington, DC.

Sprat, T. (1958). *History of the Royal Society*. In J. I. Cope & H. W. Jones (Eds.). St. Louis, MO: Washington University Studies.

Stoddard, E. W. (1985). The role of ethos in the theory of technical writing. *The Technical Writing Teacher*, **11**, 229–241.

Weisman, H. M. (1980). *Basic technical writing* (4th ed.). Columbus, OH: Merrill.

Whalen, T. (1982). *Preparing contract-winning proposals*. New York: Pilot Books.

Whalen, T. (1985). Renewal: Writing the incumbent proposal. *IEEE Transactions on Professional Communication*, **PC-28**, 13–15.

Whitburn, M. D. with Davis, M., Higgins, S., Oates, L., & Spurgeon, K. (1978). The plain style in scientific and technical writing. *Journal of Technical Writing and Communication*, **8**, 349–57.

Young, R. E., Becker, A. L., & Pike, K. L. (1970). *Rhetoric: Discovery and change*. New York: Harcourt, Brace & World.

Zappen, J. P. (1983). A rhetoric for research in sciences and technologies. In P. V. Anderson, R. J. Brockman, & C. R. Miller (Eds.), *Essays in technical and scientific communication: Research, theory, and practice* (pp. 123–138). Farmingdale, NY: Baywood.

4

How Can Rhetorical Theory Help Writers Create Readable Documents?

ROBERT DE BEAUGRANDE

Text linguist Robert de Beaugrande demystifies a persistent debate in technical writing: the relationship between logic, the content and coherence of a text, and rhetoric, a text's style of presentation. Many technical writers believe that their documents should emphasize objective fact, information, and a plain style of presentation. They view rhetoric as deliberate manipulation of unnecessary, even obfuscating, ornamentation. Thus, rhetoric becomes an attempt either to set up a figurative smokescreen or to establish credibility for a discipline by inventing a pseudo-scientific terminology.

De Beaugrande argues that technical writers who discredit rhetoric fail to understand that content and presentation, logic and rhetoric, are inseparable. The effective use of rhetoric enhances the logic of a document because rhetoric identifies strategies writers can use to engage readers. Rhetorical strategies, unlike the outmoded rules many technical writers admire, guide writers so that the choices they make as they write reflect their knowledge of the document's audience, its context and content, and the intention and purpose of individual words, sentences and paragraphs. As de Beaugrande demonstrates, all logically consistent, coherent texts adhere to the criteria for rhetorically sound texts as well.

Demystifying Rhetoric

Traditionally, an opposition has been drawn between *logic* as the content and coherence of a text versus *rhetoric* as the style of presentation. But this supposed opposition is more a matter of degree. All language is rhetorical in the sense that rhetoric is a means for substituting (using "tropes") and arranging (using "schemes"); after all, words are not the things they represent, nor is the order of words the same as the order of things, so that using language necessarily involves substituting and arranging. Conversely, all language is logical in the sense that logic is a means for organizing content and making it coherent in terms of consistency, consequences, and conclusions.

Yet in the history of ideas in western culture, logic and rhetoric are opposing concepts with the former more highly esteemed than the latter. One of our most pervasive myths has been that an objective, knowledgeable person creates text with no adornments, alternations, or obvious personal interest. Logical thought and language were held to be the norm, the ideal mode of cognition and expression. Periodically, scholars would undertake to write a "grammar" for a language in which logic was strongly emphasized and rhetoric neglected. Chomsky's (1957, 1965) "transformational grammar" is among the most recent and best known of this longstanding project. And its failure as a description of human language is emblematic for the fact that people say or write what they do for all sorts of reasons beyond the abstract structures of phrases.

It is hardly surprising that during the ascendency of industrialism and positivism, rhetoric was seldom a public concern. It faded out of the schools; in colleges, it lived on in departments of speech, barely acknowledged by departments of English. In everyday parlance, "rhetoric" became a term for flowery, insincere diction. As De Man (1979) remarks, rhetoric was regarded with suspicion as "a tool" for "the manipulation of the self and others" (p. 173) and as a "fraudulent grammar used in oratory" (p. 130). The truthful person should have no need to persuade or adorn, but need only let the facts speak for themselves. The figural use of language was consistently regarded as inferior or less desirable than the literal use.

No doubt the general hope was for a whole society of plain talkers and plain writers, where all people say briefly and clearly exactly what they mean. Instead we have a remarkably inarticulate society that, never having acquired explicit skills in rhetoric, routinely falls prey to egregious political and commercial manipulation. The majority of Americans do not consider themselves good writers, and their writing is often neither plain nor clear. As examples 1 to 3 show, writers who do not understand the resources of language often write clumsy, possibly ambiguous locutions that detract more than they add.

1. Stiff fines have also been put into effect to deter people from combining alcohol and the driver's seat.

2. Alaska has the highest mountains in the U.S. that practically leap out at you.

3. Relations with college administrators have left a bitter taste in the mouths of students for a long while.

Without a concern for figural usage, writers eventually encounter situations where literal usage cannot adequately express their ideas.

Relevant Research

The reappearance of rhetoric in the English curriculum in recent years is a hopeful sign. So far, however, there is no general consensus about how it should be practiced or taught. Currently, technical writing has not been explicitly regrounded in rhetoric. What is needed is a large-scale attempt to unite rhetoric and logic in a general, disciplined communicative awareness.

On the professional level, the urgent need of standards for improving the rhetorical and logical aspects of technical writing is readily evident. The long-standing impression that many experts write unclear and uncommunicative prose has been confirmed by a number of recent studies of how public documents are designed (e.g. Bond, Hayes, & Flower, 1980; Holland & Redish, 1981; Swarts, Flower, & Hayes, 1980). If functional documents are uncommunicative, the general public is kept ignorant of technological issues, and even experts waste valuable time and effort deciphering each other's reports. Though it would be hard to calculate, the net loss to our society in terms of information and cost is staggering.

Admittedly, the foundations for the kind of educational and professional discipline I propose are far from established. For many people, "good writing" is simply what conforms to "good grammar" and follows the conventions of spelling and punctuation. This focus has certainly been encouraged by the old-fashioned teaching of composition centered on "errors" rather than on communicative effectiveness. Theoretical research has similarly neglected the correlation between specific writing decisions and their audience impact:

> We have, in studies of form, largely a record of search for formulas and patterns in discourse, and a record of advice on the properties that well-ordered discourse ought, in the *a priori* judgement of theorists, to exhibit. But the reasons for the effectiveness of different patterns, the ways in which their parts interact, the most useful techniques of deciding upon particular sequences of steps in composing—in short, many of the fundamental topics one has to address in choosing a form for a composition—have been dealt with slightly, hesitantly or not at all. (Larson, 1976, p. 71)

This disproportion between a search for patterns and a recognition of the communicative elements of text is closely tied to a prescriptive approach to teaching: student writers are not encouraged to use their own judgment but to follow an incompletely realized set of teacher's "rules." These writers never learn to assume the perspective of the audience and to estimate the rhetorical effects of their own writing decisions. In their later careers as technical writers, they do not know what strategies they need for the more important tasks of prose organization.

This deficiency will be alleviated if writers can evaluate and revise what they write by consulting reliable decision-making criteria (see de Beaugrande, 1984a, 1984b). Writers no longer absorb rhetorical skills from merely reading the traditional fare of essays, reports, and literary works. Efficient writing can only be attained by large numbers of people if the criteria skilled writers intuitively know and apply are rendered in terms of explicit procedures that virtually anyone can follow.

In general, our society is quite naive about writing. People are both anxious and opinionated regarding what they believe to be good or bad, right or wrong. Technical writing at least has the special status of being largely carried out by people who are genuinely knowledgeable about the subject matter. But this can be a mixed blessing. Insiders all too readily judge their writing as understandable because they are so well in control of the content; they fail to read from the perspective of someone else. Also, technical jargon becomes too comfortable, and far more difficult terms are deployed than the occasion genuinely demands.

A Rhetoric of Technical Writing

The concept of "logic" described on pages 79–80 implies that rhetoric has little to do with technical writing. In fact, however, audience effects are a major concern (Mathes & Stevenson, 1976; Pearsall, 1969). More than most modes of writing, technical writing is explicitly purposive, more audience-directed than self-directed.

Technical writing is popularly believed to be distinguished by a complex, latinate vocabulary (Hogben, 1970). In reality, this belief is misleading (Rapoport, 1963) and often leads writers to decorate their technical documents with such a vocabulary whether the occasion calls for it or not. Writers will not only use specialized terms, genuinely functional because they have been carefully defined for the technical field being discussed; they will also introduce pompous, empty circumlocutions under the notion that the whole text will seem more scientific and authoritative. For example, (4a), by trying to seem more impressive, is less communicative than (4b).

> **4a.** These problems, together with the potential importance of the findings, underline the importance of similar but expanded studies to insure the gener-

alizability of the findings and more closely controlled experimental studies to isolate more definitively the important elements. (*Journalism Quarterly,* Summer 1970, p. 308)

4b. These problems urgently call for more studies and experiments to find out how general our findings are and which elements are the important ones.

Perhaps too, the writer of (4a) wanted to conceal his or her doubts about *importance of the findings* behind a linguistic smoke screen. The use of overly technical style, as in (5), certainly can be a rhetorical strategy for confusing nonexperts.

5. "Residence employee" means an employee of an insured person while performing duties arising out of and in the course of employment in connection with the maintenance or use of the residence premises, including similar duties elsewhere, not in connection with the business of an insured person.

This definition from an insurance policy is inaccessible even to expert readers, let alone to the average customer.

A somewhat more complex version of this phenomenon of combining ostentatiousness with obfuscation is found in disciplines that attempt to establish, by rhetorical means, the technicality or scientific credibility of their precepts. Since a statement such as (6a) would be transparently vulnerable, the writer chose to present his thesis in the "technicalizing rhetoric" of (6b):

6a. Our theory works on the assumption that the meaning of a sentence is understood by adding up the meanings of its words and phrases.

6b. For the semantic component to reconstruct the principles underlying the speaker's semantic competence, the rules of the semantic component must simulate the operation of these principles by projecting representations of the meaning of higher-level constituents from representations of the meaning of the lower-level constituents that comprise them. (Katz, 1966, p. 152)

As we see in example 6, emerging technical fields, such as linguistics, education, and social sciences, are trying to establish an audible modern terminology. Although their plight is understandable, a rhetoric of obscurity is a poor substitute for real insight.

Thus, the rhetoric of technical writing should be the strategic use of only those elements and resources that clarify and communicate rather than confuse. For example, one useful device is metaphors that relate a technical concept to an everyday one.

7. Waveguide systems are often loosely called "plumbing." The name implies a network of empty pipes where electrical energy flows unimpeded. Actually, a waveguide is a precisely-designed, electrically-tuned structure for propagating electromagnetic waves. It transmits certain determinable frequencies well, does not transmit some frequencies at all, and transmits others only with large losses. (Houp & Pearsall, 1973, p. 9)

The metaphor is not misleading because the differences between *wave-guide systems* and *plumbing* are immediately explained. However, technical and scientific writers are often shy about using metaphors, presumably in fear of damaging their stance of "objectivity."

It should be plain by now that technical writing cannot be contrasted to essay or literary writing by drawing a parallel contrast between logic and rhetoric. A style dictated by logic alone, even if it could be developed, would exclude many strategies that are needed to make technical writing genuinely readable. Instead, technical writers should acknowledge that their writing is also rhetorically engaged and that they can expressly and consciously practice the strategies that make their writing communicative.

Recommendations

Oddly enough, technical writers are more familiar with the less helpful notion of writing according to *rules* than they are with the notion of writing according to *strategies*. Few technical writers view themselves as consciously making informed judgments about the size and shape of sentences or paragraphs in functional documents. Their decision making is largely kept at the intuitive or unconscious level, such as it is in spontaneous speaking. The situation is much the same for revising and editing, when decision-making criteria would be especially tactical. Writers often fail to appreciate that they can treat any given content or intention with a range of options that are more or less appropriate to that context or intention. Typically, these writers view the phrasing of a passage as fixed for good once it has been committed to paper or allow only minimal changes.

Sentence Structure

Suppose you wanted to answer the question, "How long and complicated should my sentences be?" Advice such as "don't be too long or too short" or "vary between long and short" is rhetorically ineffectual because it ignores context and purpose. More useful advice is that "shorter sentences are better if the material is likely to seem unfamiliar or difficult; longer sentences are better if the material is likely to seem familiar or easy." (See Chapter 6 for further discussion.) The following examples clearly illustrate this advice. Example 8a is easy to read, even though it's a long and complicated sentence (32) words. Example 9a is hard to read, even though it's shorter (30 words).

8a. That bright light you see in the western sky right after sunset, long before any stars are visible, is Venus—not a plane, a balloon, or a UFO, as is often thought. (*Science Digest,* June 1983)

9a. An advanced air-to-air missile has intercepted a high-speed target, showing its ability to find low-flying targets amid high clutter caused by the missile's own radar returns reflecting from the ground. (*Science Digest,* June 1983)

Most people would be familiar with *seeing a bright light in the western sky right after sunset* and may have *often thought* it might be *a plane, a balloon, or a UFO.* They may have heard somewhere that the light is *Venus* though they may have forgotten it. Such background knowledge makes it easy to read a long, complicated sentence like (8a).

In contrast, most people are far less likely to know about missile interception and radar jamming. We can make (9a) easier by using shorter, simpler sentences:

9b. An advanced air-to-air missile has intercepted a high-speed target. The missile showed its ability to find low-flying targets amid high clutter. This interference is caused by the missile's own radar returns reflecting from the ground.

This pattern enhances readability by giving the materials a bit at a time. Readers can digest the message in installments.

On the other hand, we don't gain much by using shorter, simpler sentences for example 8a. The result is merely tedious:

8b. You see a bright light in the western sky. You see it right after sunset. You see it long before any stars are visible. That light is Venus. It is not a plane, a balloon, or a UFO. People have often thought so, though.

To develop your awareness of such considerations, you can practice revisions during which sentences are either broken down or merged. For instance, one writer rewrote (10a) and (11a) as (10b) and (11b):

10a. Exercises increase flexibility of the muscles, so they should be done, because unless patients get exercise, they will be confined to a wheelchair sooner than is necessary.

10b. Exercises increase flexibility of the muscles. Patients should get exercise in order not to be confined to a wheelchair sooner than is necessary.

11a. We took 3 grams of the compound. We had it in a test tube. We stirred it for 10 minutes. We kept it over a small flame.

11b. We stirred 3 grams of the compound in a test tube over a small flame for 10 minutes.

This training is helpful for learning to see sentence structure as flexible and to adapt sentence structure according to the level of the audience.

Jargon

Another issue I raised before is how to decide when specialized vocabulary is useful and when it is merely ornamental. The latter often serves to conceal trite or disagreeable content. For example, (12a) and (13a) give an illusion of being more presentable than (12b) and (13b):

12a. An accumulation of time periods has as its consequence an increase in the human age factor.

12b. As time goes by, people get older.

13a. Our pacification initiative adversely affected civilian female and juvenile segments of the population.

13b. Our bombing killed civilian women and children.

You can distinguish between such ornamental versus the justified use of the scientific and technical terms of your field by revising passages so as to retain only necessary technical jargon, e.g., making (14a) into (14b):

14a. Various forms of behavioral disturbances result from the children's inability to participate in communication.

14b. The children's behavior is disturbed if they can't communicate.

You can also revise to make technical passages more readable for a general audience. You can explain difficult terms and supply background knowledge. For instance, you could rework an example like (15a) into something like (15b):

15a. Electrophoretic separation techniques are hindered by gravity-induced convection mixing. (*Science Digest,* June 1983)

15b. Electrophoresis is a means of using electrical fields to separate a mixture of molecules. Gravity can hinder this electrophoretic separation by making some molecules flow other ways besides those in the electric field, so that mixing occurs.

Nothing is more antagonizing than an inaccessible text whose writer presupposes too much specialized knowledge. Readers should not have to run to the library to understand your technical prose.

These illustrations show what kinds of practice a technical writer might do in order to develop rhetorical awareness. You should relate your decisions to considerations of readability and audience impact.

A Test Case in Rhetoric

You may well be wondering whether you can carry this training in rhetoric over to real-life situations when you have a report to write. You may be thinking that your intuitive, unconscious writing habits are pretty well set as they are, no matter what you might practice. This objection is an important one, and skepticism certainly is in order. Training does not easily transfer to everyday practice of largely automatic skills.

In principle, practice in terms of general strategies is more likely to carry over than practice in terms of special "rules" (like "never begin a sentence with 'and' "). Strategies are positive and productive, whereas rules are negative and inhibiting. Also, these strategies apply to revising, so you don't have to interpose them directly in the on-line creation of the first draft. Thus, when you revise, your potential for conscious intervention is much better.

I will close with a test case of a writer who did in fact acquire significant skills at revising during a semester of training. He was given the following paper as an in-class test. I edited it to incorporate the typical problems I found in students' writing: the paper rambles, repeats itself, gets things in no clear order, seems hardly aware of purpose, and marks no paragraphs.

> **16a.** The state of the modern world of today is one that is not very encouraging, if you know what I mean. Matter of fact, it's more than a little alarming it's very alarming. There being several reasons. First, one of those reasons is the reason that raw materials are getting scarce and there isn't as much of them around as what their used to be once upon a time. The countries that still produce raw materials wanting more adequate payment, there people are fairly poor. After that, the next thing that happens is that the industrial nations with the industries have to raise prices for finished goods. Inflation resulting all across the market. Everything costs more, wages have to rise as well, eventually the price of raw materials must be raised again. Well, all this being hard on the economy of small nations who depend heavily on international trade with other nations they trade with. Another reason the spread of advanced weapon systems. Almost any country can afford dangerous weapons that they might use for any occasion when they think they might win something or gain an advantage over some other country, we see what can happen from what's happening in the Middle East right now. A very explosive situation, needless to say; And in a few years many, more countries will have atomic warheads that could go off by accident or could be casually used for petty little things that don't matter much, you wouldnt think. But before it can be stopped, large-scale annihilation and people getting killed everywhere. Another reason having to do with the emergence of many complicated power blocks. Not just the U.S. and the U.S.S.R. with their allies facing each other like two football teams before the kick-off. Many smaller but still powerful) blocks with a variety of intrests in world politics. All requiring careful diplomicy. Its harder than ever to predict how the seen might change and what results might result from an international event affecting many contries. Such as the shooting down of the Korean airliner. It is certain that there can be no doubt of that fact that in the coming years of the future that's ahead of us we'll be seeing and witnessing the most, profound changes in the world situation in all

history. A time of great anxiety but also of great hope. That's the mood of the world going toward the end of the twentieth century and into the twenty-first.

The student had about 50 minutes to write a completely new version, taking care of as many barriers to communication as possible. No aids were allowed except a dictionary.

16b. THE WORLD TODAY

The state of the world of today isn't very encouraging. The scarcity of raw materials, the spread of advanced weapon systems, and the emergence of complicated power blocks have all combined to form an extremely alarming situation.

Raw materials are becoming rare. Because they are poor, the countries that produce raw materials are now asking higher prices for their precious goods. The industrial nations must, in turn, raise their prices for finished goods.

As a result, there is widespread inflation. To keep up with the cost of living, wages must rise as well. Eventually the price of raw materials will be hiked again. This inflation is very hard on small nations, who's economy depends heavily on international trade.

Another danger in the world today is the spread of advanced weapon systems. Most countries can now afford these dangerous weapons. These instruments of destruction could be used at any time by a country that thinks they might have something to win or can gain an advantage over another country (as is evident in the Middle East right now). Needless to say, it's a very tense situation.

In a few years, many countries will have atomic warheads. These bombs might explode by accident or be casually used to kill. Before an accidental war could be stopped, large-scale destruction will have taken place.

The emergence of complicated power blocs is also a problem that confronts the world. The U.S. and the U.S.S.R. with their allies—small but still powerful—face each other like two football teams before kick-off. Both countries have a variety of interests in world politics, requiring careful diplomacy on both of their parts. Its harder than ever to predict the results of an international event (such as the shooting down of the Korean airliner).

As you can see, the world today is not very encouraging. There is a shortage of raw materials, a spread of new weapons, and an emergence of power blocs. There's no doubt that in the coming years we'll be seeing the most profound changes in all of history. As the twentieth century comes to a close, the world shares a feeling of greater anxiety, but also of much hope.

Though still not great prose, the paper communicates the writer's thoughts more clearly because it is rhetorically much better controlled than the original was. The writer has made the choices that will encourage an audience to read fluently and follow the line of argument readily.

I shall mention only a few of the useful changes the student made. He recognized and eliminated most of the phrases with little content (*if you know what I mean, matter of fact,* etc.). He also didn't bother to state things that obviously followed from what was being said, e.g., that *casual use* is directed to *petty little things*. He cut out the obvious redundancies, such as that *international trade* has to be done with *other nations they trade with*. When he did refer back to prior content, he would vary the expressions: *getting scarce . . . becoming rare . . . asking higher prices . . . raise prices . . . hike prices . . . weapons . . . instruments of destruction;* and *atomic warheads . . . bombs*. Pronouns with vague content were eliminated, such as the original *they* for *country*, or were specified, such as *it . . . accidental war*.

The main points were both previewed in the opening paragraph and reviewed in the concluding paragraph, more tersely in the latter than in the former. Paragraphs were marked off in the middle according to these main points, and transition statements were used to look back explicitly to the main topic: *Another danger in the world today is . . . ; ". . . is also a problem that confronts the world*. Clearly, the writer not only attended to the mechanical problems but also was able to deal with rhetorical organization.

Obviously, this brief demonstration does not cover all rhetorical aspects that could be relevant to the more advanced technical writer; many more still need to be accounted for. It merely demonstrates that a rhetorically conceived program for writer training can clarify and guide the concrete performance. You can certainly profit by becoming more aware of your rhetorical strategies and by practicing them as you write. Ideally, you may see with the eyes of your audience well enough to be genuinely accommodating and considerate. Writing is a buyer's market, and you should give your customers the best.

References

de Beaugrande, R. (1984a). *Text production: Toward a science of composition.* Norwood, NJ: Ablex.

de Beaugrande, R. (1984b). *Writing step by step.* New York: Harcourt Brace Jovanovich.

Bond, S., Hayes, J. R., & Flower, L. (1980). *Translating the law into common language.* Technical Report 8. Pittsburgh, PA: Carnegie-Mellon Document Design Center.

Chomsky, N. (1957). *Syntactic structures.* The Hague: Mouton Press.

Chomsky, N. (1965). *Aspects of the theory of syntax.* Cambridge, MA: M.I.T. Press.

de Man, P. (1979). *Allegories of reading.* New Haven, CT: Yale University Press.

Hogben, L. (1970). *The vocabulary of science.* New York: Stein & Day.

Holland, V. M., & Redish, J. (1981). *Strategies for understanding forms and other public documents*. Technical Report 13. Washington, DC: Document Design Center.

Houp, K., & Pearsall, T. (1973). *Reporting technical information*. Beverly Hills, CA: Glencoe.

Katz, J. (1966). *The philosophy of language*. New York: Harper & Row.

Larson, R. (1976). Structure and form in non-fiction prose. In G. Tate (Ed.), *Teaching composition* (pp. 45–71). Fort Worth: Texas Christian University Press.

Mathes, J. C., & Sevenson, D. (1976). *Designing technical reports*. Indianapolis, IN: Bobbs-Merrill.

Pearsall, T. (1969). *Audience analysis for technical writing*. Beverly Hills, CA: Glencoe.

Rapoport, A. (1963). The language of science. In M. Blickle & M. Passe (Eds.), *Readings for technical writers* (pp. 20–32). New York: Ronald.

Swarts, H., Flower, L., & Hayes, J. R. (1980). *How headings in documents can mislead readers*. Technical Report 9. Pittsburgh, PA: Carnegie-Mellon Document Design Center.

5

How Can Functional Sentence Perspective Help Technical Writers Compose Readable Documents?

JOSEPH M. WILLIAMS

Information presented logically to an audience is a goal of every technical writer. However, readers often find functional documents difficult to understand because they cannot follow the flow of information in the documents. The topic of a document, what the report is about, may be obscured; the information in the document, the writer's commentary on the topic, may seem confusing. This problem is a rhetorical one best addressed by a systematic analysis of where information is placed both in sentences and in groups of sentences.

Professor Joseph M. Williams of the University of Chicago illustrates how a rhetorical strategy known as Functional Sentence Perspective (FSP) allows writers to arrange information systematically. As a principle, Functional Sentence Perspective parallels the grammatical distinction that separates the subject of a sentence, what the sentence is about, from the predicate, what comments develop the subject. However, the process by which writers give their audiences contexts for ideas goes beyond a grammatical analysis. FSP generalizes that writers should, whenever possible, prepare their readers for new information by beginning their sentences with a "topic," ideas that are familiar to the audience or that have already been referred to, and then moving to the "comment": newer, less predictable, less familiar information.

By consistently choosing to arrange information this way, writers create topic strings that enhance the coherence of their documents or that emphasize one concept over another.

Functional Sentence Perspective as a Principle of Style

Consider the following passage:

> **1a.** Alteration of mucosal and vascular permeability through elaboration of a toxin by the vibrio is one current hypothesis in explanation of this kind of severe dehydration. **2a.** Changes in small capillaries located near the basal surface of the epithelial cells and the appearance of numerous microvesicles in the cytoplasm of the mucosal cells is evidence in favor of this hypothesis. **3a.** Hydrodynamic transport of fluid into the interstitial tissue and then through the mucosa into the lumen of the gut is believed to depend on capillary permeability alteration.

If we apply the familiar advice about editing abstract nominalizations into verbs, we can improve this passage considerably (Williams, 1984):

> **1b.** More permeable mucosal and vascular tissue results when a toxin is elaborated by the vibrio, according to one current hypothesis that would explain this kind of severe dehydration. **2b.** That small capillaries located near the basal surface of the epithelial cells change and that numerous microvesicles appear in the cytoplasm of the mucosal cells is evidence in favor of this hypothesis. **3b.** Fluid hydrodynamically transported into the interstitial tissue through the mucosa into the lumen of the gut results when capillaries become more permeable.

We have changed all the abstract nominalizations to verbs, but the passage is still less than entirely readable. If we edit this passage one more time in some common sense ways, we create a version that is considerably more readable.

> **1c.** We can explain this kind of severe dehydration by the hypothesis that the vibrio elaborates a toxin that makes mucosal tissue more permeable. **2c.** In favor of this hypothesis are changes in the small capillaries located near the basal surface of the epithelial cells and the appearance of numerous microvesicles in the cytoplasm of the mucosal cells. **3c.** We believe that when capillaries become more permeable, fluid is hydrodynamically transported into the interstitial tissue and then through the mucosa into the lumen of the gut.

We have renominalized two verbs: *change* and *appear,* and we have introduced the researcher/writer in the form of *we,* a change that some find inappropriate for scientific writing. But we have made in some ways a more substantive change as well. It is true that a style characterized by verbs is more readable than a style characterized by nominalizations derived from those verbs. But there is a hierarchically higher principle of style that goes beyond straightforward translation of nouns into verbs: it is the principle of Functional Sentence Perspective (FSP).

In brief, if that principle can be stated as a rough rule of thumb, it would sound like this: arrange the order of information in a sentence so that you begin with the information that is repeated, more familiar to the reader, more psychologically "accessible," less psychologically complex, less important, and move toward information that is newer, less familiar, less psychologically accessible, and more complex and important.

Relevant Research

This principle of style has long been a staple of rhetoricians who have recognized the structural importance of climax. But it was not articulated as a linguistic principle until the 19th century (Weil, 1887), and it was not until the 20th century that linguists, particularly in Eastern Europe, began to take a serious interest in the matter (Danes, 1974; Enkvist, 1973; Enkvist & Kohonen, 1976; Firbas, 1964, 1968, 1974; Halliday, 1974; Li, 1976; Mathesius, 1928).

These criteria are not synonymous; indeed, sometimes they conflict. Sometimes the writer must find a way to compromise among varying demands that call for different orders of information. Occasionally, the writer will decide that something already referred to is the most important unit; occasionally, the most important element will be shorter and less complex than what precedes it. And in some cases, the grammar simply does not allow a writer to arrange the order of ideas in the order that might seem best. But in general, the criteria coincide: what is less complex is usually more familiar; what is more important is usually less predictable.

If we look again at the differences between the two passages, we can recognize how comparable sentences in the second passage systematically move from less familiar to more, from less predictable to more predictable, and from more complex to less complex. Compare (1b) and (1c). In (1b), the sentence opens with the crucial information. It is this unit that explains what has been mentioned in a previous sentence as the problem: *this kind of severe dehydration,* a phrase that appears not at the beginning of the sentence, where repeated information should appear, but at the end. In the *when* clause, the subject is *a toxin,* an indefinite phrase communicating unpredictable information; following the passive verb is a *by* phrase referring to a concept that anyone familiar with cellular biology would understand: *the vibrio,* the definite *the* signaling presupposed knowledge. The reference to *one current hypothesis* refers to a concept that a reader of a medical journal would fully expect to find; medical writing predictably discusses hypotheses. But the sentence refers to this relatively predictable concept at the end of the sentence, in conjunction with information that, as we said, repeats what has already been mentioned: *this kind of severe dehydration.* In other words, sentence 1b runs backward, from what is relatively newer and unpredictable to what is older, familiar, or reasonably presupposed.

In (1c), on the other hand, the information systematically moves from the most presupposed, familiar, accessible to the least: it begins with *we,* a reference to the authors/researchers whose names appear under the title of the article. The researchers predictably explain things, particularly problems they have announced earlier: *this kind of severe dehydration.* The reference to hypothesis sets up the following new information: *the* vibrio, a reference to information a reader would control, elaborates *a* toxin, relatively newer information, but still not surprising since that is what vibrio (a kind of bacteria) commonly do. This toxin then makes mucosal and vascular tissue *more permeable,* the point of the sentence, the explanation, the most important, the least predictable information.

We need not track the differences between (2b)–(2c) and (3b)–(3c); they follow precisely the same pattern: in (1b) to (3b), newer, less familiar, less accessible, unpredictable information fits with the least surprising information last; in (1c) to (3c), older, more familiar, more accessible, presupposed information comes first; newer, less familiar, less accessible, unpredictable information later, with the most important information last.

FSP as a Bilevel System

We can formalize this principle of style as shown in Figure 5.1.

Topic	Comment
Theme	Rheme

Figure 5.1 FSP as a Bilevel System

By a *Topic* we mean that syntactic unit that constitutes the psychological subject. Ordinarily, this syntactic unit is also the grammatical subject and at least canonically expresses older, more accessible information (though as we have seen, it does not always do so). In the sentence,

4. These larger issues I will address presently.

the grammatical subject is, of course, *I.* But the psychological subject, the *Topic* is *these larger issues,* the preposed, or "topicalized" direct object of *address.* It is what the sentence is "about," in the sense of what the writer wants to elaborate on, explain, discuss, comment on. English provides few overt signals of nonsubject *Topics:*

5. *In regard to* these larger issues, there is little to say.

6. *As for* China, who can predict what will happen?

7. *Turning now to* isomorphic sets, we can see that . . . .

This issue of *Topic* is, however, more complex than first meets the eye. In some cases, a reader cannot determine from the surface of the sentence alone what the *Topic* is but must interpret it from the context. The *Topic* of this next sentence could be either *1933* or *the United States:*

8. In 1933, the United States experienced the Great Depression.

If the writer has set out to write about everything that happened in 1933—the rise of Nazism in Germany, famine in China, etc.—then the reader will take 1933 as the governing topic of the sentence:

9. Nineteen thirty-three was one of the terrible years in world history. That year saw the rise of Nazism in Germany. It was the year of the great famine in China. In 1933, the United States experienced . . . .

But if the writer is obviously discussing various things that happened to the United States through history, then the governing *Topic* is *the United States:*

10. America has experienced many troubles in this century. In 1933, the U.S. experienced the Great Depression. In 1941, the U.S. was drawn into WWII. In 1951, we found ourselves enmeshed in the Korean Conflict. In 1965, we became involved in Viet Nam. In 1972, ...

In this context, then, the phrase *in 1933* merely locates, orients, the reader toward what follows.

It should be noted that *Topics* organize our understanding of not just whole sentences but of clauses and even complex phrases as well:

11. When the retaining pin is displaced, the connecting rod will disengage, an event occurring with some regularity in earlier models.

In the opening clause of (11), the *Topic* is *the retaining pin;* in the main clause, the topic is *the connecting rod*. In the phrase *an event occurring with some regularity in earlier models,* we can identify *an event* as a subtopic. In other words, a sentence may consist of a network of *Topics* arranged in a hierarchical order. In this next sentence, the prepositional phrase announces the governing *Topic* for a unit of discourse *(formal schemata)* and the subject announces the local *Topic* for the sentence *(the crucial role of interactive networks):*

12. In regard to formal schemata, the crucial role of interactive networks influences any model we might attempt to construct.

In short, the *Topic* of a sentence is defined simultaneously by position and by function: it is usually one of the first noun phrases, but it must be the psychological unit on which the writer intends to elaborate.

For all practical purposes, we can define *Comment* as whatever is not *Topic*. If we wished to be more exact, we would have to distinguish among other functional elements in a sentence. One such element is *metadiscourse*, that language that refers not to the primary subject matter of the discourse but to the act of discoursing (Williams, 1984). This paragraph is largely metadiscourse. If we eliminated everything but the primary subject matter, we would have this left:

> **13.** *comment* . . . whatever is not *Topic*. . . . *other* functional elements in a sentence. One such element is "metadiscourse," that language which refers not to the primary subject matter of the discourse but to the act of discoursing. This paragraph is . . . metadiscourse. . . . primary subject matter . . . this. . . .

Metadiscourse allows us to comment on what we are saying, to refer to the reader, to the organization of the discourse, and so on.

There are other functional elements. We've already mentioned *orienters*, those introductory phrases that locate a reader in place and time. There are also *attributors*, usually in the form of a subject and verb:

> **14a.** Donaldson claims that there exists a direct correlation between GNP and rate of savings. This correlation does not exist, however, when. . . .

But it may also be in the form of a prepositional phrase:

> **14b.** According to Donaldson, there exists a . . . .

An attributor merely gets information on the table in an authoritative way. It may be better defined as a species of metadiscourse, along with elements such as *we think*, when they only provide a measure of truth-value:

> **14c.** This correlation, we think, does not exist when . . . .

This unit consisting of *Topic–Comment* is a fixed, structural unit. It is completed by the *Theme–Rheme* unit, a unit that is variable and semantic. *Theme* is that information that we have described as older, presupposed, that which the writer assumes the reader will feel is psychologically more accessible. *Rheme* is that information that is relatively less accessible. Because this concept of "relatively more/less" is conceptually a gradient, we ordinarily can find no strict demarcation line between *Theme* and *Rheme*.

It is important to note that *Theme–Rheme* is not a syntagmatically fixed level of analysis. Unlike *Topic* and *Comment*, which are an ordered pair, *Thematic* and *Rhematic* information may appear anywhere in a sentence.

This description differs from most of the literature on this subject by distinguishing *Topic* from *Theme* and *Comment* from *Rheme*. In much of the literature, the terms are often used inconsistently, often interchangeably (Barri, 1978; Dahl, 1976; Firbas, 1964, 1966; Li, 1976; Perfetti & Goldman, 1975; Sgall & Janicova, 1979a, 1979b). But as we saw in the opening set of examples, *Theme* (old) and *Rheme* (new) can be permuted through *Topics* and *Comments* and so cannot be conflated with *Topic* and *Comment*:

> **15a.** *Increased vascular permeability* will explain this kind of severe dehydration.

> **15b.** *This kind of severe dehydration* can be explained by increased vascular permeability.

The Psychological Reality of FSP

It has been established that the order of *Theme–Rheme* is the most "readable" order of information (Haviland & Clark, 1974; Kieras, 1981; Vande Kopple, 1982; Williams, 1979, 1984; Witte, 1983; Witte & Faigley, 1981). Among competent writers, it is the most frequently observed order. It is for this reason that most researchers have conflated *Theme* with *Topic*: since thematic information canonically appears early and since *Topics* by definition appear early, it was easy to assume some rule-like correspondence. But as we have seen, the "rule" can be regularly "broken" by writers entirely oblivious to any canonical order.

It is important to note how the *Topic–Comment/Theme–Rheme* unit corresponds with another bilevel principle of style we illustrated in the difference between the opening pair of excerpts. In revising the a sentences into the b sentences, we tried to observe this principle of style, which is shown in Figure 5.2.

Subject	Verb	Complement
Agent	Action	——

Figure 5.2 FSP as a Bilevel System

This is a principle of style that synthesizes the relatively fixed surface order of subject–verb–complement with the underlying case relationships developed in case grammars (Fillmore, 1968). Case grammars stipulate many more case relationships than merely agency, but for our purposes, this cruder representation will serve. By *Agent* we ordinarily mean the source of an *Action* or condition.

On the basis of a good deal of research on the matter, we may generalize that sentences whose subjects express the *Agency* of an *Action* will read more clearly than those that do not (Clark, 1965; Freeman, 1978; Hutten-locher, Eisenberg, & Strauss, 1968; Johnson, 1967; Lippman, 1972; Slobin, 1968). On the basis of this and other research along the same lines, we can reasonably conclude that, just as a *Theme/Topic–Rheme/Comment* is the reader's canonical order, so is *Agent/Subject*.

And we can make the same claim about *Verbs* and *Actions*: a considerable body of research has argued persuasively that prose in which *Actions* correspond with *Verbs* is cognitively less taxing than prose in which *Actions* are systematically displaced through nominalizations (Coleman, 1964, 1965; Williams, 1979, 1984). Our own intuition about complexity and cognitive demands should suggest that sentence b below is more "readable" than sentence a:

> **16a.** Our discussion as to the termination of the program was followed by a decision as to staff reduction.
>
> **16b.** After we discussed whether to terminate the program, we decided to reduce the staff.

We might now point out the considerable power of a model of sentence structure that integrates the two levels of a grammar based on syntactic and case roles with the two levels of an FSP grammar based on rhetorical and informational roles, which is shown in Figure 5.3.

Topic		Comment
Theme		Rheme
Subject	Verb	Complement
Agent	Action	——

Figure 5.3 Integrated FSP/Syntactic Model

The power of this model is that it entails a wide range of stylistic principles beyond those we have already described:

1. *Short subjects:* If the topic/subject is an agent of the action, it will usually be familiar; therefore, once the cast of characters in a discourse has been introduced, the agents of actions will be familiar, hence referable with short, unelaborated phrases (Yngve, 1960).
2. *Long complements:* Since new information is usually syntactically more complex than old, the more complex information will appear at the end of the sentence.
3. *Topic strings:* The consistent topicalization of repeated elements will create a consistent "topic string."

Topic Strings

A *Topic String* is a sequence of *Topics* through a series of clauses and/or sentences that constitutes a coherent set of concepts. If we look at the *Topics* in the first three passages we analyzed at the beginning of this chapter, we can contrast the inconsistency of the *Topics* in the first two with the relatively more consistent *Topics* of the third:

1a. Alteration of mucosal . . by the vibrio.

2a. Changes in small capillaries . . . mucosal cells.

3a. Hydrodynamic transport . . . lumen of the gut.

1b. More permeable mucosal and vascular tissue . . . .
>   a toxin
>   a hypothesis

2b. That small capillaries . . . mucosal cells.

3b. Fluid hydrodynamically transported . . . the gut capillaries . . . .

1c. We (can explain)
>   hypothesis
>   vibrio

2c. We (believe)
>   capillaries

In the a and b versions, the *Topic/Subjects* are long and complex, almost entirely unpredictable, certainly inconsistent. In c, the *Topics* are all short and simple. In the c sentences, the first level of *Topics* all have to do with hypotheses, explanations, etc. The second level of *Topics* are all drawn from information mentioned at least once in the previous discourse. The *Topics* constitute a coherent *Topic String*, the basis for part of that epiphenomenon of text we call "coherence" (Vande Kopple, 1982; Witte, 1983; Witte & Faigley, 1981). This sense of coherence, form, structure arises out of the interplay of several layers of perception. It depends both on what we systematically know about the subject matter of the text, the "scheme" or "frame" we bring to our reading of the material (Rumelhart, 1980), and on our ability to perceive the rhetorical superstructure of the text (Van Dijk, 1980). But it also depends on an emergent sense of consistency through a series of clauses. It is this emergent sense of consistency out of a consistent *Topic String* that helps a reader construct both the rhetorical superstructure and the schema or frame of the content.

The Relative Nature of Theme and Rheme

And it is just here, in what the writer assumes about the reader's familiarity with the subject of the discourse, that we have to point out the entirely *relative* nature of *Theme–Rheme* distinctions, for though we may casually use the terms "old" and "new," "familiar" and "unfamiliar," we have to understand that what is old or familiar to one audience will be new and unfamiliar to another. Consider these two passages:

> **17a.** An understanding of the role of calcium in the activation of muscle cells through the interaction of the contractile proteins actin and myosin, and the regulatory proteins tropomyosin and troponin, the sarcomere, the fundamental unit of muscle contraction, best provides an appreciation of the effects of calcium blockers. **17b.** The interaction of myosin, an ATPase or energy-producing protein, in its thick filament with actin in its thin filament is regulated by tropomyosin and troponin in the thin filament. **17c.** Troponin C, which binds calcium; troponin I, which participates in the actin–myosin interaction; and troponin T, which binds troponin to tropomyosin constitute three peptide chains of troponin. **17d.** An excess of 10 for the myoplasmic concentration $C++$ leads to its binding to troponin C. **17e.** The inhibitory forces of tropomyosin on actin are removed and its complex interaction with myosin is manifested as contraction.

> **18a.** If we can understand how the contraction of muscles depends on calcium, we can appreciate how muscle cells are affected by calcium blockers. **18b.** The fundamental unit of muscle contraction is the sarcomere. **18c.** The sarcomere has two filaments, one thick and one thin. **18d.** They contain proteins that cause contraction and proteins that prevent it. **18e.** The thick filament contains one protein, myosin. **18f.** It is an ATPase, or energy-producing protein. **18g.** The thin filament contains three proteins: one is actin, which interacts with myosin in the thick filament to cause contraction. **18h.** The other two are proteins that inhibit contraction: they are tropomyosin and troponin. **18i.** The troponin consists of three chains of peptides: troponin I; troponin T, which binds troponin I to tropomyosin; and troponin C, which binds calcium. **18j.** When a muscle relaxes, calcium binds to tropomyosin in the thin filament. **18k.** This inhibits the actin in the thin filament from interacting with the myosin in the thick filament. **18l.** A muscle contracts when the myoplasmic concentration $C++$ in the sarcomere exceeds 10. **18m.** At that point, the calcium bound to the tropomyosin in the thick filament releases and binds to troponin C. **18n.** The tropomyosin then no longer inhibits the interaction of actin in the thin filament and myosin in the thick filament, and the muscle contracts.

If you are unfamiliar with the cellular mechanics of calcium blocking, then the second passage is far easier to read than the first. One immediately obvious reason is that the first compresses a great deal of information into a few long sentences; the second unpacks those sentences and breaks out the information into smaller bundles. But the issue here is not the mere *length* of the sentences: rather it is the consequences of identifying what a reader ignorant of this subject would take to be entirely new information and then *deliberately locating that new information at the end of clauses so that the NEXT sentence can topicalize the now old, thematic informa-*

tion in preparation for the next bit of new information. The second version pushes to the end of each of its clauses the more complex information, i.e., information that may feel baffling because of its terminology or because of the complexity of its content. In the first version, each clause is front-loaded with complex information—at least from the point of view of the naive reader.

In fact, when medical personnel familiar with the subject have read these passages, they have reported that the first is "pretty well written," "good for medical prose," etc., whereas the second is "simpleminded," "written for dummies." The reason for this response is obvious: for those knowledgeable readers, there is very little new information here. In fact, it is virtually all old information.

For many writers, the fact that a sarcomere has a thick and a thin filament is not obvious. It is obvious to a researcher familiar with these matters. It is information as familiar as that implied in the space between these two sentences:

> **19.** I enjoy fishing. My old bamboo rod has been a friend to me for so long that I don't even remember where it came from.

Is there anyone who does not know that part of fishing *could* include an old bamboo rod? Probably not. Is there anyone who does not know that sarcomeres have filaments, that there are two of them, that they are thick and thin, that they consist of proteins? Just about everyone reading this chapter.

The point is this: old and new are entirely relative terms, not just to each other, but *to the reader as well.* If this is the case, then the writer has to think very carefully about how to lay out new terms; how to arrange complex bundles of information; how to make explicit what might otherwise be left to inference. If we are writing for an audience that is relatively unfamiliar with the subject matter, then we must deliberately attend to the process of *putting unfamiliar terminology and complex concepts at the ends of sentences,* of introducing that terminology and those concepts with syntactic elements that serve *only to contextualize that unfamiliar information.* That device allowed us to rewrite the first calcium blocker passage into the second. The original sentence is:

> **17b.** The interaction of myosin, an ATPase or energy-producing protein, in its thick filament with actin in the thin filament is regulated by tropomyosin and troponin in the thin filament.

In this revised passage, I have put in parentheses information implied in example 18 and capitalized what I would take to be rhematic (i.e., new) information.

18c. (The sarcomere has) TWO FILAMENTS, one THICK and one THIN. **18d.** They contain PROTEINS (THAT CAUSE CONTRACTION AND PROTEINS THAT INHIBIT it). **18e.** The thick filament contains one protein, MYOSIN. **18f.** It is an ATPase, or ENERGY PRODUCING protein. **18g.** The thin filament contains THREE proteins: one is ACTIN, which INTERACTS WITH MYOSIN in the thick filament (to CAUSE CONTRACTION). **18h.** The other two (are proteins that INHIBIT CONTRACTION. (They) are TROPOMYOSIN AND TROPONIN.

If we are concerned with communicating complex information effectively and efficiently, we ought not be reluctant to lay out that information in ways that may seem to some inordinately simple. The key to organizing that information is to feed the reader new information a bit at a time, always contextualized by preceding, older information.

Recommendations: FSP and the Nuances of Rhetorical Intentions

We have, then, a powerful generalization about the structure of sentences in context. It is a generalization that synthesizes multiple levels of sentence structure into a systematic model of style and that recommends itself to all writers who are interested in readability at levels of structure beyond single sentences. Given only this principle, those writers who struggle to make themselves syntactically clear could mechanically apply this principle and be sure that, in general, their prose would be substantially more readable.

But those writers who want to be not only clear but pointed will have to go beyond this mechanical application. We must note that in some discourse, the cast of characters is either irrelevant or nonexistent, or at least distinctly secondary to a topic more conceptually based. In this next passage, for example, the topics are consistent but not drawn from the cast of characters.

> **20a.** *Chinese* is a difficult language to learn to read. **20b.** *Its written form* consists of many thousands of characters that are based not on phonetic forms but on concepts. **20c.** Thus *a character* communicates not a sound but an idea. **20d.** *This kind of ideographic script* requires immense amounts of time to memorize and has contributed greatly to the immense gulf between the intellectual and nonintellectual classes in Chinese history.

The discourse topic is written Chinese, so the *Topics* are, variously, *Chinese, its written form, a character, this kind of ideographic script*. Thus, the concept of "cast of characters" must be expanded to include those unitary, nameable concepts that "act" as conceptual components in a larger concept.

And this now brings us to one more major complication in the theory of FSP. I have been describing FSP and its relation to the syntactic/case structure of sentences as if we could describe the patterns of old and new, *Topic* and *Comment,* almost by some set of cut-and-dried rules that would invariably lead us to the canonical, the most desirable form of a sentence: old first, new last (depending on the knowledge of the reader, of course), agents as subjects, actions as verbs. Although that generalization has considerable power, it is too simple. In fact, it is in the writer's power to *make* old that which will constitute the topics of the following sentences. It is in the writer's power to *choose* from among the cast of characters, concrete or conceptual, those entities that will best serve as topics.

Consider this revision of the passage on Chinese writing:

> **21a.** *A speaker of Chinese* can learn to read his language only with great difficulty. **21b.** *The reader* must deal with many thousands of characters that are based not on phonetic forms, but on concepts. **21c.** *The reader* must devote immense amounts of time to memorizing this kind of ideographic script. **21d.** As a result, *the Chinese intellectual* class has always been separated from the nonintellectual by an immense gulf.

This passage is pragmatically identical to the first version, but instead of a consistent set of *Topics* centering on the concept of the Chinese writing system, its *Topics* center on *Agents,* on Chinese speakers and writers. I suspect that when you read the first version, the one consistently topicalizing aspects of the writing system, it seemed appropriate (or at least did not seem obviously inappropriate) that the *Topics should* have centered not on speakers and writers but on elements of the writing system.

My point is that we must recognize this major complication of any simple theory of FSP. Because in a complex universe of reference we have many identifiable semantic elements that we *could* topicalize, we must *choose* which set of them we *should* topicalize (Penelope, 1982; Shopen & Williams, 1981; Williams, 1981). As a default condition, we might argue that we should choose animate agents first. If we find that syntactically or pragmatically inconvenient (because we have no animate agents in our field of reference), then we should topicalize whatever concept we would define as the topic of the whole discourse. I have made the *Topics* of *this* paragraph *we,* forcing into the *Comment* of each sentence the more complex conceptual material. Indeed, I have self-consciously written this whole chapter in a relatively *Agent–Action* kind of style, partly out of personal habit, partly because it is the easiest to read. And since this subject matter is not entirely transparent on first reading, it requires a prose style as clear as I can conveniently manage. But I did not have to topicalize any *Agents* at all. I could have written this entire paragraph without referring to writers or to me:

> **22a.** The point here is a major complication in the theory of FSP: because a complex universe of reference makes available many identifiable semantic ele-

ments that could be topicalized, it is necessary to choose just those that *should* serve as topics. **22b.** As a default condition, it might be argued that animate agents should take priority. **22c.** If that choice is syntactically or pragmatically inconvenient (because no animate agents exist in the field of reference), then the sentence should topicalize whatever concept seems to be the topic of the whole discourse.

This passage feels less personal, more academic. Assuming that a writer *chooses* voice, tone, rhetorical stance in order to achieve a particular intention, the choice of topics in this version would be a consequence of higher order choices.

But even the default condition, i.e., choosing one of the cast of characters as a consistent topic, fails as a simple principle when we recognize that most subject matters contain X number of characters, any one of which could be topicalized. For example, here are three versions of the same passage from a legal brief analyzing liability in regard to patients who develop unfavorable reactions to prescribed drugs. The first is unfocused; it has no consistent topic string. But if we rewrite it to make one of the cast of characters the topic, we have to choose *which* character, the doctor or the patient.

23a. *Unmonitored reactions* may also lead to physician's liability. **23b.** In *M. v. L., Spandoline* had been prescribed, which resulted in agranulocytosis. **23c.** *The manufacturer's literature* indicated the need for frequent observations, and any sudden development of infection immediately reported. **23d.** Furthermore, *no white cell counts* were made until after the development of a sore throat. **23e.** Also, *evidence* indicated that *no instructions* had been given to report any signs of agranulocytosis. **23f.** Even though *a laboratory* was the defendant, *the court* held for negligence on the part of the physician.

Before you read the next two versions, you might decide whether the patient or the doctor should be topicalized.

24a. If *a physician* fails to monitor a patient's reactions, *he* may also be found liable. **24b.** In *M. v. L., a physician* prescribed Spandoline, which resulted in the patient developing agranulocytosis. **24c.** *The physician* had been cautioned by the manufacturer's literature that *he* should observe the patient frequently and immediately report any signs of infection. **24d.** Furthermore, *the physician* made no white cell counts until *the patient* developed a sore throat. **24e.** Also *the physician* failed to provide evidence that *he* instructed the patient to report any signs of agranulocytosis. **24f.** Even though *the physician* was not the defendant but rather a laboratory, *the physician* was held negligent.

25a. If *a patient* is not monitored by his physician, *he* may successfully charge the physician with negligence. **25b.** In *M. v. L., a patient* who had been taking the prescribed drug Spandoline developed agranulocytosis. **25c.** According to the manufacturer's literature, *the patient* should undergo frequent observation and should report any sign of infection immediately. **25d.** Furthermore, *the patient* had no white cell count taken until after *he* developed a sore throat.

25e. Also *the patient* showed that *he* had never been instructed to report any signs of agranulocytosis. 25f. Even though *the patient* brought the action against a laboratory, *the court* held the physician to be negligent.

If we assume that the object of such a passage is to persuade a judge or jury to find in favor of the patient, then the writer's job is to assign responsibility for all damaging actions to the physician, to make the physician the source, the *Agent*, the responsible entity. But more than that, the writer also wants to make the physician *the center of attention*, the *Topic* of the discourse. And if that is so, then the best way to write each sentence is to make the physician the consistent *Agent/Subject/Theme/Topic*.

Thus the issue of selecting *Topics* and *Thematic* information cannot be separated from issues involving hierarchically more important factors, particularly that of *audience and intention*. Every writer must solve, simultaneously, a host of problems involving intention, audience, his or her own *persona*, selection of genre and form, particular microstructural issues of style, including everything we have been discussing here. But these elements do not constitute a mere bundle of elements. They are hierarchically arranged. Governing all is intention. Intention defines the writer's stance toward the *persona* to project, toward what *persona* to project onto the audience. And all of this determines the particular genre the writer will adopt, and the genre delimits the range of possible forms the writer may select. And the particular form the writer wants will determine the kinds of choices made at the level of particular sentence structure.

We process discourse top–down and bottom–up simultaneously. We have to decode at the level of letters and punctuation, but we interpret in the context of perceived intention manifested in a text in a particular genre, all informed by assumptions about implied form. Thus, although we have discussed Functional Sentence Perspective as a phenomenon whose context seems to be bounded by the forms of sequences of sentences, we must recognize that the larger context—especially intention—is always central; until those more fundamental issues are decided, a writer cannot determine the appropriate way to manifest the crucially important principles of Functional Sentence Perspective.

References

Barri, N. (1978). Theme and rheme as immediate constituents. *Folia Linguistica*, 12, 253–265.
Clark, H. (1965). Some structural properties of simple active and passive sentences. *Journal of Verbal Learning and Verbal Behavior*, 4, 365–370.

Coleman, E. (1964). The comprehensibility of several grammatical transformations. *Journal of Applied Psychology,* **48,** 186–190.

Coleman, E. (1965). Learning of prose written in four grammatical transfigurations. *Journal of Applied Psychology.* **49,** 332–341.

Dahl, O. (1976). What is new information? In N. Enkvist & V. Kohonen (Eds.), *Symposium on the interaction of parameters affecting word order: Report on text linguistics* (pp. 37–50). Abo, Finland: The Research Institute of the Abo Akademi Foundation.

Danes, F. (1974). Functional sentence perspective and the organization of the text. In F. Danes (Ed.), *Papers on functional sentence perspective* (pp. 217–222). The Hague: Mouton.

Enkvist, N. (1973). Theme dynamics and style: An experiment. *Studia Anglica Posnaniensia,* **5,** 127–135.

Enkvist, N., & Kohonen, V. (Eds.). (1976). *Report on text linguistics.* Abo, Finland: The Research Institute of the Abo Akademi Foundation.

Fillmore, C. (1968). The case for the case. In E. Bach & R. T. Harms (Eds.). *Universals in linguistic theory* (pp. 1–68). New York: Holt, Rinehart & Winston.

Firbas, J. (1964). On defining theme in functional sentence analysis. *Travaux linguistiques de Prague,* **1,** 267–280.

Firbas, J. (1966). Non-thematic subjects in contemporary English. *Travaux Linguistiques de Prague,* **2,** 239–256. Tuscaloosa: University of Alabama Press.

Firbas, J. (1974). Some aspects of the Czechoslovak approach to problems of functional sentence perspective. In F. Danes (Ed.), *Papers on functional sentence perspective* (pp. 11–37). The Hague: Mouton.

Freeman, C. (1978). Readability and text structure: A view from linguistics. In *Final report to the Carnegie Corporation, children's functional language and education in the earl years.* Arlington, VA: Center for Applied Linguistics.

Halliday, M.A.K. (1974). The place of functional sentence perspective in the system of linguistic description. In F. Danes (Ed.), *Papers on functional sentence perspective* (pp. 43–53). The Hague: Mouton.

Haviland, S.E., & Clark, H.H. (1974). What's new?: Acquiring new information as a process in comprehension. *Journal of Verbal Learning and Verbal Behavior,* **13,** 512–521.

Huttenlocher, J., Eisenberg, K., & Strauss, S. (1968). Comprehension: Relation between perceived actor and logical subject. *Journal of Verbal Learning and Verbal Behavior,* **7,** 527–530.

Johnson, M. (1967). Syntactic position and rated meaning. *Journal of Verbal Learning and Verbal Behavior,* **6,** 240–246.

Kieras, D. (1981). Topicalization effects in cued recall of technical prose. *Memory & Cognition,* **9,** 541–549.

Li, C. (Ed.). (1976). *Subject and topic.* New York: Academic Press.

Lippman, M. (1972). The influence of grammatical transform in a syllogistic reasoning task. *Journal of Verbal Learning and Verbal Behavior,* **11,** 424–430.

Mathesius, V. (1928). On linguistic characterology with illustrations from modern English. In J. Vachek (Ed.), *A Prague school reader in linguistics* (pp. 398–412). Bloomington: Indiana University Press.

Penelope, J. (1982). Topicalization: The rhetorical strategies it serves and the interpretive strategies it imposes. *Linguistics,* **20,** 683–695.

Perfetti, C., & Goldman, S. (1975). Discourse functions of thematization and topicalization. *Journal of Psycholinguistic Research*, **4**, 252–271.

Rumelhart, D. (1980). Schemata: A framework for text understanding. In R. Spiro, B. Bruce, & W. Brewer (Eds.), *Theoretical issues in reading comprehension*. Hillsdale, NJ: Erlbaum.

Sgall, P. & Janicova, E. (1979a). Toward a definition of focus and topic: Part I. *Prague Bulletin of Mathematical Linguistics*, **31**, 3–25.

Sgall, P., & Janicova, E. (1979b). Toward a definition of focus and topic: Part II. *Prague Bulletin of Mathematical Linguistics*, **32**, 24–32.

Shopen, T., & Williams, J. M. (1981). *Style and variables in English*. Boston: Winthrop.

Slobin, D. (1968). Recall of full and truncated passive sentences in connected discourse. *Journal of Verbal Learning and Verbal Behavior*, **7**, 876–881.

Vande Kopple, W. (1982). Functional sentence perspective, composition, and reading. *College Composition and Communication*, **33**, 50–63.

Van Dijk, T. (1980). *Macrostructures*. Hillsdale: NJ: Erlbaum.

Weil, H. (1887). *The order of words in the ancient languages compared with that of the modern languages* (C.W. Super, Trans.). Boston: Ginn.

Williams, J. M. (1979). Defining complexity. *College English*, **40**, 595–609.

Williams, J. M. (1981). Literary style: The personal voice. In T. Shopen & J. Williams (Eds.), *Style and variables in English* (pp. 117–216). Boston: Winthrop.

Williams, J. M. (1984). *Style: Ten lessons in clarity and grace* (2nd ed.). Glenview, IL: Scott Foresman.

Witte, S. (1983). Topical structure and revision. *College Composition and Communication*, **34**, 313–341.

Witte, S., & Faigley, L. (1981). Coherence, cohesion, and writing quality. *College Composition and Communication*, **32**, 189–204.

Yngve, V. (1960). A model and a hypothesis for language structure. *Proceedings of the American Philosophical Society*, **104**, 444–466.

6

How Can Functional Documents Be Made More Cohesive and Coherent?

LYNN BEENE

[I]t does not matter how one gets the effect of ease [in writing]. For my part, if I get it at all, it is only by strenuous effort. Nature seldom provides me with the word, the turn of phrase, that is appropriate without being farfetched or commonplace.

Somerset Maugham, *Mr. Maugham Himself*

Technical writing, like all prose, depends upon the logical ordering of ideas in a text and the obvious signaling of those ideas. Professor Lynn Beene of the University of New Mexico argues that writers can create readable, informative, and persuasive prose by applying the principles of cohesion and coherence recursively to their drafts. Cohesion, such as anaphoric and cataphoric reference, gives solidity and fluidity to individual sentence patterns in a text. Cohesive markers allow writers to signal relationships among ideas in a document. In everyday language, markers illustrate five basic kinds of connection: reference, substitution, ellipses, lexical reiteration and collocation, and conjunction. However, cohesion is not limited to reference items or transitional words and phrases; the heart of cohesion is found in the ways writers use clauses to reintroduce important topics.

Coherence, unlike cohesion, describes the larger organizational structures in documents: coherent structures ensure the document's overall communicativeness and may not be obviously encoded in the text. Coherent documents not only use cohesive markers, they also present information in logical patterns that link two broad levels of knowledge: world knowledge, information readers bring to a document, and text-presented knowledge, information writers present in a document.

Given this background information and theory, Beene recommends that writers use two basic text strategies: one for informational patterns and one for cohesive structures. By managing information, writers can shape documents to suit an audience and can systematically direct the flow of information so that readers can easily assimilate important data. Beene discusses the role of several key elements in the production of cohesive and coherent texts including grammatical structure, word choice, conjunction, sentence patterns, voice, and paragraphing. Writers can use these key textual elements to achieve cohesion by focusing upon surface components in a document, and writers can achieve coherence by manipulating logical relationships among sentences.

Like Maugham, technical writers can make documents cohesive and coherent by structuring the information carefully, by evaluating the prose critically, and by revising a draft conscientiously. However, technical writers frequently use time pressures as reasons to avoid this needed evaluation and revision. Or they may feel that functional prose, unlike fiction or poetry, doesn't warrant revision. Neither excuse is acceptable. Employers who have to muddle through incoherent prose will find it necessary either to have the writer explain the document or to have it rewritten. Either way, the employer loses time and money. Further, technical writers, per se, are not given the gift other writers wish for: the ability to produce clear, readable first drafts. In good functional writing, everything "hangs together" because the writer makes it hang together. In other words, communicative technical writing, documents that are both cohesive and coherent, come about the same way Maugham created his texts—by strenuous effort.

Although technical writers are concerned with these textual elements, few of us understand how the relationships of cohesion and coherence govern text. We know that some texts are more communicative than others. In some documents the words, phrases, and sentences fit the ideas better; the level of formality is appropriate to the audience; and the groups of sentences fall together as if they were written for the readers' easy assimilation. We also know that in less communicative texts the ideas are obscured by the words or the arrangement of sentences; the language is either too formal, too informal, too technical, too general, or vacillates among these extremes; or there is no logical relationship among the sentences. What we need to understand are the dynamics that make texts communicative and how to use these dynamics to benefit our writing.

Technical documents communicate because they are *cohesive* and *coherent,* complex concepts that are difficult to define and identify. At their most basic, cohesion and coherence are different levels of the same phenomenon: the logical ordering of ideas in texts. Whereas cohesion identifies the specific items such as articles, nominals, conjunctions, or sentence patterns writers use to connect sentences to one another, coherence describes the strategies and structures writers use to distribute information and to communicate ideas in an orderly manner. Despite their different levels of analyses, cohesion and coherence are generally recognized as basic elements a document must have if it is to be readable, informative, and persuasive.

This chapter defines and illustrates the concepts of cohesion and coherence in ways that are specifically helpful for technical writers. Then the chapter summarizes the most useful theoretical positions on cohesion and coherence while illustrating how cohesion and coherence are most often achieved in technical documents. Finally the chapter details how writers can revise their documents to make them more cohesive, coherent, and, ultimately, more communicative.

Relevant Research

Cohesion

Writers signal the relationships among ideas in a document by using certain words or grammatical structures as cohesive markers. Cohesive markers, such as pronouns or conjunctions, direct readers to interpret a part of one sentence by referring either to a word or phrase in a previous sentence (anaphoric reference) or to a word or phrase in a subsequent sentence (cataphoric reference). In (1), for example, readers must go back to the first sentence in the sequence to find the implied meaning for the word *this* (i.e., a reduced risk of radiation detection and exploitation).

> 1. Because the airborn radar needs to be turned on only intermittently, the risk is reduced that its radiation will be detected and exploited by an enemy. *This* makes it possible for the first time to consider true automation of the terrain avoidance function. (*Aviation Week and Space Technology,* July 1982, p. 73)

Anaphoric reference is frequently used in texts that elaborate or summarize concepts. With cataphoric relationships, however, readers must keep one idea in mind until they find the full explanation of it, or complete reference for it, in the succeeding text. Thus cataphoric reference is more common in instructions.

Although anaphora and cataphora are basic relationships identifiable in any connected discourse, researchers disagree on precisely what means writers use to encode these relationships in their documents. Some theorists believe that speakers and writers use the interaction of syntax and grammar, in identifiable patterns, to reveal connections in their discourse (de Beaugrande & Dressler, 1981). Others argue that the roles that sentences play within paragraphs (e.g., statement, restatement, qualification, evaluation; Larson, 1967) or the possible interactions (e.g., coordination, conclusion) among sentences in a paragraph (Winterowd, 1970) explain how discourse is connected.

Despite such theoretical debates, anaphoric and cataphoric relationships are best understood by examining how they are realized in everyday English texts. Halliday and Hasan (1976) classify cohesive markers into five basic categories of connectors: reference, substitution, ellipsis, lexical reiteration and collocation, and conjunction. According to Halliday and Hasan, these categories are so ingrained in language that they are, for the most part, unnoticed. For example, reference ties include pronouns (e.g., *it, he, them, she*), demonstrative/determiners (e.g., *this, that, those, the*), and comparatives (e.g., *more, bigger, less*). If we analyze examples such as (1), we see just how reference ties can be overlooked in a document. In (1) the pronoun *it* appears as both a specific referent, *its* [the radar's] radiation, and a general referent, the phrase *makes it* [the situation] *possible*. As readers we take little notice of the specificity that the demonstrative *the* implies in *the airborn radar* or *the terrain avoidance function*, but we are lead to wonder at exactly who the enemy is because of the writer's use of *an* instead of *the*.

Of all reference items, comparatives are the most typically anaphoric and have the greatest variety. Comparatives identify quality (e.g., *bigger, biggest*), quantity (e.g., *more, less, as many as*), difference (e.g., *other, additional*), or similarity (e.g., *like, such*) (Halliday & Hasan, 1976). In example 2 comparatives of quality *(substantiality)*, quantity *(more vulnerable)*, and difference *(than)* tie the two sentences together by explaining why tightly controlled timing is vital.

> 2. Tight controlled timing is also necessary to execute a successful launch-under-attack policy. For the first three or four minutes of flight US ICM's would be *substantially* more *vulnerable* to nuclear explosions over their base areas *than* they would be in their silos. (*Scientific American*, Jan. 1984, pp. 38–40)

Substitution, Halliday and Hasan's second catagory, is a stylistic form of cohesion that replaces one word with its synonym and allows a writer to avoid redundancy. Science writers often substitute nouns (e.g., *one* for *number* in example 3), verbs (e.g., *does* for *rises* in example 4), or clauses (e.g., example 5).

> 3. New numbers popped out of the equation in orderly fashion. An electron in an atom has a principal quantum number n (somewhat similar to the n in the Bohr theory), a second one 1 that restricts the size of the angular momentum of the electron's orbit around the nucleus and a third one that restricts the orientation of that angular momentum. (*Scientific American*, Jan. 1984, p. 142)

4. The effect of an inversion is to concentrate the sun's morning heat into a relatively shallow layer of air, which is why the temperature can rise very rapidly in the morning. As it *does,* the surface of the earth (but not the sea) warms, which is when the thermals and the clouds begin to appear. (*Sail,* May 1984, p. 64)

5. We think of the presentation of science as a series of propositions, one after another, as deriving from the mathematics of Euclid. *And so it does.* (*The Ascent of Man,* J. Bronowski, p. 233)

Because readers of technical reports read for information and expect precise, unambiguous statements, they will accept substitution more easily than ellipsis, substitution by zero. Ellipsis, as in example 6, slows readers down by making them read a document more closely to reconstruct the writer's point.

6. Where did Mendel get the model of an all-or-nothing heredity? I think I know [where Mendel got this model], but of course I cannot look into his head either. (*The Ascent of Man,* J. Bronowski, p. 387)

Reference, substitution, and ellipsis are less complex means of creating cohesion than using lexical ties or conjunctions. Lexical ties include reiteration, cohesion by repetition or use of synonyms or near-synonyms (e.g., *self-contained pioneering communities* for *space cities* in example 7), and collocation, cohesion by maintaining a continuity of word choice (e.g., example 3)

7. *Space cities* would be a kind of America in the skies. They also would greatly enhance the survival potential of the human species. But the project is extremely expensive, costing at minimum about the same as one Vietnam war (in resources, not in lives). In addition, the idea has the worrisome overtones of abandoning the problems of the Earth—where, after all, *self-contained pioneering communities* can be established at much less cost. (*Broca's Brain,* C. Sagan, p. 37)

Reiteration, in many instances, is simply a form of substitution. And collocation is nearly always present in technical documents because technical fields use unique, specialized vocabularies.

The most linguistically complex means of creating cohesion is conjoining because conjunctions, unlike the previous catagories of cohesive devices, may "reach out" in any direction (and probably to any distance) in a document to "express certain meanings which presuppose the presence of other components in the discourse" (Halliday & Hasan, 1976, pp. 226–227). Conjunctions impose a writer's point of view on a document. Thus a writer who sees a causal connection between ideas is likely to make this connection obvious by using causal conjunction (e.g., *consequently, therefore*).

Conjunctions are classified according to the four main relationships they establish among clauses: *additive, temporal, adversative,* and *causal.* The additive relation, expressed with conjunctions such as *and, nor, further-more, for example,* or *moreover,* always imposes connections between ideas by a strategy common in spoken English: expansion by addition (e.g., example 8).

> 8. The American anthropologist Franz Boaz . . . made short work of the fabled cranial index by showing that it varied widely both among adults of a single group and within the life of an individual. . . . *Moreover,* he found significant differences in cranial index between immigrant parents and their American-born children. (*The Mismeasure of Man,* S. J. Gould, p. 108)

The temporal relationship, expressed with conjunctions such as *then, fi-nally,* or *first . . . last,* parallels the additive one in that it expresses development by accretion. However, unlike additive relationships, with temporal conjunctions writers have to structure their additions chronologically and signal both their order and their time sequence. Commonly writers balance additional information and time sequence by combining temporal conjunctions (e.g., *first* and *next*) and a repeated sentence structure (e.g., imperative structure for instructions).

With additive and temporal relationships, events happening in a sequence of real time impose a context on documents for their writers; however, in adversative and causal relationships, writers create the sequences in the documents. The adversative relationship, expressed with conjunctions such as *yet, however, on the other hand,* or *as a matter of fact,* requires a writer to show not only a sense of sequencing in time but also to indicate how that sequence can run contrary to initial expectations. The causal relationship, expressed with conjunctions such as *thus, with this intention, hence,* or *so,* is the most complex of the four types because it imposes an additional interpretative requirement: a writer must both interpret information and let readers know that the text is an interpretation not a chronology (e.g., example 9).

> 9. If you've been assigned to do a thorough revision of a current manual, your best source of information is the audience. They know what worked for them, what didn't, and where they went wrong.
>
> *So* look through letters, reader-response forms, and any other feedback. (*How to Write a Computer Manual,* J. Price, p. 203)

Given the complications common to adversative and causal relationships, it's easy to understand why we can more quickly provide appropriate conjunctions in an instruction manual or a progress report than in an analysis or evaluation: in the former type of reports, the information sequence is already determined. However, real time is not a factor in organizing the latter type of reports.

Treating cohesion as primarily lexical and classifying cohesive markers by types (e.g., reference, ellipsis) and relationships (e.g., additive, causal) limits any evaluation of cohesion because it ignores clausal connections. Jordan (1979, pp. 2–3) argues that clause relationships, specifically those "between the referent and the clause containing the substitution," and the ways writers reintroduce or reenter topics form the heart of cohesion.

Basing his analysis on Winter's (1977) discussion of the semantic relations of sentence connectors, Jordan posits three small but quite intricate classes of clausal connectives: sentence subordinators, conjoiners, and connectors. Subordinators (e.g., *because, although, before*) introduce grammatically dependent clauses. Typically the information in these clauses is "old" material, information the writer has just presented in the document. (See Chapter 5 of this anthology for further discussion of old versus new information.) Conjoiners (e.g., *and, or, but, whereas, nor, yet, so, for*) primarily link two sentences, with little more punctuation than a comma, and indicate relationships between those two sentences such as addition (similar to additive and temporal relationships), contrast (similar to adversative), or reason (similar to causal). Connectors (e.g., *moreover, too, conversely, accordingly, finally, meanwhile*) are generally adverbs that appear at the beginning of a sentence to divide the ideas in a text into small groups. By breaking ideas in a text into small units and by providing readers with specific signals such as sentence subordinators or connectors, writers limit the amount of information readers have to follow and direct the flow of that information. In brief, they make their documents cohesive.

Jordan further argues that when writers muddle words used as connectives, their texts appear illogical (Jordan, 1974b). Consequently, if writers use connectives appropriately (i.e., to support a logically organized document), they avoid muddled logic by relating their ideas in ways best suited to both their readers' needs and the purposes and tone of their documents (Jordan, 1974a). The question, of course, is how to use these connectives "appropriately."

Jordan's (1979, p. 5) answer is commonsensical: connections between and among clauses can be defined in terms of questions "that characterize and clarify the semantics of each relation." A typical question is "How can I do X?" By asking this rhetorical question at the end of the previous paragraph, I set up presuppositions between the last sentence of that paragraph and this paragraph, namely, that this paragraph would answer the posed question. Other clausal relationships would involve questions such as (based on Winter, 1977; Jordan, 1979, 1984):

Question	Presupposing Relationship
Is X true?	Verification
Why is X so?	Analysis

If X is not so, what is so?	Verification
Why does X exist?	Speculation
Where does X go?	Location
What do I do with X?	Purpose/Conclusion
Has X been done before?	Result
Who did X?	Agency
Can X be done?	Possibility
What do I do after X	Progress/Process
What do I think of X	Evaluation

In addition to classifying cohesive connectives, Jordan notes that cohesion in writing is determined by how writers successively "reenter" a topic. To reenter main topics, writers use lexical items that either reintroduce a main topic or explore ancillary issues of that topic. For example, in this section I've discussed Jordan's emendations to a classification of cohesive markers. At several points I've "reintroduced" the main topic of discussion, cohesion, by using techniques such as repeated lexical items (e.g., *cohesion, presupposition, relations, relationships*), adverbial modifiers to summarize information (e.g., *By breaking ideas in a text into small units and by providing readers with specific signals such as connectors*), reference items (e.g., *The* temporal relationship parallels *the* additive *one* because *it* expresses . . . , *writers* limit . . . *they* make), similar sentence patterns (e.g., Subordinators . . . introduce grammatically dependent clauses . . . Conjoiners . . . primarily join two sentences . . .), and sentence connectors (e.g., *furthermore, thus, typically*). Each reintroduction is an example of reentry; each is designed to add information, in a cohesive pattern, to the base I'm building.

Jordan's two major types of reentry are "basic" and "associated." Basic reentry, similar to major topic headings in an outline, signals a continued discussion of one topic. In example 10 the main topic, the debt-restructuring agreement that defers $14 million, is reintroduced by repetition *(the $14 million, deferred, the restructuring agreement, the new agreement, the $14 million)* and pronominal reference (*it* refers to the $14 million).

10. World Airways has completed a debt-restructuring agreement that defers about $14 million in aircraft debt and lease payments. *The $14 million* would have been due next June 30. It includes $4 million that has been repeatedly *deferred* while *the restructuring agreement* was being negotiated. Under *the new agreement, the $14 million* could be repaid in 1984–1986. (*Aviation Week and Space Technology,* July 1982, p. 31)

Associated reentry, the use of part of a previously mentioned main topic as the new topic for discussion, expands a given main topic by focusing on a subtopic. Most technical writers must use both basic and associated reentry in one passage because their topics are complex. For example, (11) would have a minimal structure such as (12).

> **11.** Voyager 2's Saturn encounter began on June 5, 1981, at a distance of 77 million kilometers from Saturn. The instruments began a four-month long scrutiny of the Saturian system. . . . By July 31, when Voyager 2 was 24.7 million kilometers from Saturn the narrow-angle camera could no longer capture all Saturn's disk in a single picture. Then the camera began photographing small segments of the disk: the pictures fit together in mosaics to reveal details in the banded clouds.
>
> By August 10, even four-photograph mosaics of Saturn did not capture the entire disk. Both the wide-angle cameras were directed, instead, to concentrate on atmospheric phenomena that deserve special attention. Mosaics and single photographs covered progressively smaller sections of the planet. The infrared instrument began to map the temperatures of Saturn and the satellites. . . . (Cited in Houp & Pearsall, 1984, p. 126)

12. Voyager 2's Saturn encounter began on June 5, 1981 . . .

The instruments began a four-month long scrutiny	Associated reentry
the narrow-angle camera . . . in a single picture (By July 31)	Basic reentry
the camera began . . .	Basic reentry
the pictures fit . . . in mosaics	Associated reentry
. . . the four-photograph mosaics (By August 10)	Basic reentry
. . . wide-angle cameras . . . atmospheric	Basic reentry
. . . atmospheric phenomena . . .	Associated reentry
Mosaics and single photographs . . .	Basic reentry
The infrared instruments . . . the temperatures . . .	Basic reentry
the temperatures of Saturn . . .	Associated reentry

Like basic reentry techniques, associated reentry allows writers to use cohesive markers to link ideas and to stress certain points over other points

(Jordan, 1982). Unlike basic reentry, associated reentry introduces a related, or associated, topic rather than focusing on one main topic. Thus, when writers successfully use associated reentry as well as basic reentry techniques, they create more sophisticated, complex documents. But to use these techniques successfully, to write cohesive documents, writers must analyze their writing by anticipating what typical questions readers will have. Thus, Jordan's suggested techniques extend a lexical classification of cohesive markers to include grammatical structures and information distribution (see Chapter 5), rhetorical concerns (see Chapters 3 and 4), and audience analysis (see Chapter 2).

Coherence

When technical writers revise their documents, they don't look just for cohesive relationships. They also judge the document's overall structure and communicativeness; in other words, they evaluate its coherence as well as its cohesiveness. The question is, what is coherence?

Coherence, a more elusive and fascinating aspect of text than cohesion, has frequently been defined by reference to its results rather than its unique features. For example, 19th-century definitions stress connections between individual sentences as the keys to coherent texts (e.g., Bain, 1877). These connectives—pronouns, conjunctions, repeated words, transition words— are what contemporary linguistic theory identifies as cohesive markers. Linguists divide cohesion and coherence to emphasize an obvious point: sentences can have numerous connections between them and still not necessarily create a coherent or communicative text. Coherent documents display a larger, more comprehensive structure than cohesion identifies; they have a structure that represents the logical patterning that any writer must convey to build in the readers' minds "the same (or nearly the same) structure of ideas that the writer has in his or hers" (Flower, 1981, p. 184).

Metaphorically, coherence may be understood by reference to an analogous concept in information theory, entropy. Entropy, the measure of disorder or chaos present in any system, exists because any system, no matter how its parts are related, always has a measure of chaos present that is working against orderliness. No system has either an ideal order or total disarray; every system is a mixture with some systems having more order than chaos. Whereas chaos is an easier state than order to achieve, order is more satisfying, up to a point, because it is creative: it allows a variety of new arrangements to be created from the existing parts of a system. In other words, it is easier to have a system—be it DNA, global political alliances, or your dinner arrangements—in a mess than it is to have that system in perfect order. However, perfect order isn't an ideal state because a perfectly ordered system may become predictable and, therefore, uninformative. Ultimately, the information must be structured to meet the needs of the users and must have guidelines (i.e., signals) so its users can follow the structure. Thus, coherent texts are communicative systems that are more ordered, have less entropy, than incoherent texts.

Linguists believe that coherence involves at least two broad levels of knowledge: world knowledge and text-presented knowledge. All native speakers of a language unconsciously share "world knowledge," including beliefs, assumptions, commonsense identifications, and experiences. Generally writers need not state this information explicitly because it should be mutually understood. To organize their world knowledge, native speakers rely on conventional patterns (Schank & Abelson, 1977). These patterns, known variously as frames, schema, plans, and scripts, allow language users to organize knowledge so they can use it. Frames describe general knowledge about familiar concepts (e.g., "high school graduation"). Schema order experiences by time or causality (e.g., finding information in a library). Plans structure knowledge in terms of intended goals (e.g., a problem-solving strategy), and scripts describe courses of action that must be followed in certain situations (e.g., ordering dinner in a restaurant) (de Beaugrande & Dressler, 1981, pp. 90–91). Readers use frames, schema, plans, and scripts to reduce the mass of information in a text into a structured, coherent whole by anticipating the organization and development in any text. Thus, readers will try to "make sense" out of a document, to reduce its entropy, no matter how incoherently it may be written.

Text-presented knowledge, on the other hand, is not known to all language users until a writer structures old/given and new information in a text and triggers connections between the text-presented knowledge and the readers' world knowledge. A writer's ability to trigger these connections and the logical patternings of ideas in a document creates a text's coherent structure. Coherence, therefore, entails explanations of semantic relationships beyond those indicated in the actual words of the text.

Generally, linguistic definitions of coherence focus on how language users structure text-presented knowledge, the propositions of a text, into a linguistic system that facilitates communication. The definitions posit that, when they communicate information to readers, writers accomplish two tasks simultaneously: they predicate, or say something about something, and they promulgate, or make information publicly available. To accomplish these goals, writers try to identify and cohesively relate their propositions by creating inferences or logical relationships among propositions. Writers can relate propositions in at least two ways: they can add information by conjoining it to previous statements or they can change or contradict previous statements. The former can always be freely done; the latter must be justified in some way for readers to accept it.

Briefly, then, coherent texts are communicative systems that are more ordered than entropic, that are consistent in both the organization and signaling of information structures, and that exploit readers' world knowledge by linking that knowledge to information in a document. In other words, ideal coherence texts

1. Use readers' knowledge of similar systems (i.e., texts).
2. Order information.
3. Avoid inconsistency.
4. Balance the old and new information.
5. Reveal relationships by using cues (i.e., cohesive markers).

Incoherent texts violate some or all of the general principles listed above. In incoherent texts, for example, grammatical errors distract readers and prevent them from integrating the new information in the document (the text-presented knowledge) with what they already know. A writer may fail to state a topic clearly, inconsistently organize the main topics or subtopics, or omit signaling devices needed to clarify relationships among propositions. Any or all of these situations can defeat a text's coherence. More often, texts appear incoherent because the information flow has been interrupted for some reason. Frequently, this interruption comes from the writer misjudging the amount of information readers have or need. If the writer assumes the readers have more knowledge about a topic than they actually do, the text will be obscure. If the writer assumes less knowledge, the text will be uninformative because it is predictable.

In sum, defining coherence precisely is a far more difficult task than defining cohesion despite the similarity between the two linguistic concepts. Where cohesion identifies specific lexical features or grammatical relationships that tie sentences together, coherence identifies textual patterns and cognitive procedures that make texts communicative and relevant to readers. Thus coherence goes beyond cohesion, encompassing not only sentence-to-sentence connections and transitions among paragraphs but also propositions and their inferences, information structures and their distribution, and access routes that connect readers' world knowledge to text-presented knowledge. All of this is done in an effort to reduce disorder and make readers understand complex information.

Recommendations

Although some technical writers organize information carefully before writing and, thus, avoid extensive revision, more often they create functional documents by pulling together bits and pieces of information that had no inherent organizational structure and then revising the draft until it's readable. Making a first draft into a coherent and cohesive document depends primarily on who will be reading the document and why. But once a writer knows something about a document's audience and purpose, that writer can make the various pieces of information flow smoothly by using two basic text strategies: one for informational patterns and a second for cohesive structure. These two strategies promote coherence by encouraging a writer to create overlapping presuppositions and to repeat lexical items (Markels, 1984).

Information Management

To enhance a document's coherence, writers should use their knowledge of their audience to examine how the information in the document is distributed. Consider two points about the audience. First, how much information do the readers need? Second, what information can be omitted? Information readers already know is appropriate in the background section of a document; information readers need to know to take action on a document should obviously dictate the document's overall organization. Information of interest primarily to the writer is inappropriate and must be omitted. Two useful techniques are to put important information both in discursive and graphic form (see Chapter 7) and to relegate anything not central to the document's purpose to the appendixes.

However, the information management is not just a matter of adding and deleting certain facts: writers must also systematically direct the flow of information. To do this, writers need to place old/given information (i.e., what readers know or what is being written about in a sequence of sentences) before new information (i.e., knowledge the writer adds to what readers know). Most readers unconsciously expect old information to appear in a sentence-initial position and new information in a sentence-final one (see Chapter 5). To promote readers' understanding, writers can move information around by, for instance, altering syntactic patterns (e.g., changing sentences from active voice to passive voice). More often, writers use introductory clauses to summarize information from a prior sentence.

Information distribution is entwined with lexical and referential cohesion, syntactic patterning, and a standard rhetorical strategy: state the old before the new. In (13) the writers use all three of these text-creating features to describe collecting vessels. Notice that every sentence in this passage begins with a reference to these vessels (i.e., lexical and referential cohesion). This repetition identifies the vessels as the topic of the passage, the old information, and puts the emphasis on the identifying features, the new information.

> **13.** Lymphatic vessels of the first order arise in the lymphatic capillary plexus which they drain. They all enter lymph nodes as afferent vessels. From these nodes, efferent vessels usually pass to one or to a series of lymph nodes before they join the thoracic duct or the right lymphatic duct. The larger collecting vessels often extend over long distances without change of caliber. The lymphatic vessels are exceedingly delicate, and their coats are so transparent that the fluid they contain is readily seen through them. They are interrupted at intervals by constrictions, which correspond to the situations of valves in their interior and gives them a knotted or beaded appearance. (*Anatomy of the Human Body,* H. Gray and C.M. Gross, p. 739)

Elements of Cohesion

The next revision, that for cohesion, focuses on surface components in a text (e.g., word endings, conjunctions, lexical items, sentence patterns, paragraph structure) and on how these items show the logical relationships among sentences. Cohesive elements are not merely added to make a technical document pleasant to read; they guide the document's readers through the writer's information system analogous to the way a cataloguing system guides a library's users through its information. As the following brief overview of the practical application of cohesive elements illustrates, in a functional document nearly every word contributes to cohesion.

A basic and frequently overlooked aspect of cohesion is grammatical structure. If the grammar is flawed, readers have more difficulty linking sentences and integrating the details of the text into a coherent structure. They are forced to figure out the correct grammar, and that effort diverts their attention from the document's information. Checking such obvious points as subject–verb agreement is only the beginning. A consistent sequence of verb tenses, for example, is a little noted but important cohesive technique. *Consistent sequence* means that in a sequence of sentences the actions that occur at the same time appear in the same tense. In (4) the author's use of present tense verbs (i.e., verbs that indicate an action contemporaneous with the sentence) helps not only to tie the sentences in the paragraph together but also to clarify the ellipical adverbial clause, *as it does,* used to summarize information from the previous sentence.

> **4.** The effect of an inversion is to concentrate the sun's morning heat into a relatively shallow layer of air, which is why the temperature can rise very rapidly in the morning. *As it does,* the surface of the earth (but not the sea) warms, which is when the thermals and the clouds begin to appear. (*Sail,* May 1984, p. 64).

Appropriate word choice, or the *level* of formality in a document, depends upon the primary readership of that document. For example, technical jargon of a specific field is a form of shorthand for an audience knowledgeable about the field. In brief, it is part of the world knowledge some readers bring to certain functional documents. However, specialists in other disciplines and general readers will not understand undefined or unexplained specialized terms and will view the text as opaque. These readers will need definitions and explanations, knowledge introduced to them by the text. Likewise, abbreviations should be defined when they are first used; once identified, they are anaphorically cohesive.

Using a word in a prior sentence to anticipate an explanatory assertion in subsequent clauses is an effective form of cataphoric cohesion. In (14) *message* forces reader to infer a connection between it and the information in the subsequent sentence, the behavior of orchids. No other connection is needed; the readers have done the work by inference.

14. The *message* is paradoxical but profound. Orchids manufacture their intricate devices from the common components of ordinary flowers, parts usually fitted for very different functions. (*The Panda's Thumb*, S.J. Gould, p. 20)

A few of the most common anticipatory nouns in technical writing are

accident	diagnosis	point
accomplishment	difference	precaution
achievement	difficulty	problem
act	disadvantage	process
advantage	discovery	proof
agreement	effect	proposal
aim	event	proposition
alternative	example	purpose
answer	fact	qualification
aspect	factor	quality
asset	goal	question
benefit	implication	reason
cause	importance	similarity
change	improvement	statement
characteristic	incident	step
choice	information	strategy
conclusion	message	suggestion
condition	method	technique
consequence	observation	theory
criterion	occurrence	understanding
danger	outcome	weakness
decision	plan	

Conjunction and lexical coherence are among the most important and complex items in technical writing. Conjunctions, the more obvious cohesive markers, identify semantic relations such as additive *(and)*, temporal *(then)*, adversative *(yet)*, and causal *(so)* (Halliday & Hasan, 1976). Although conjunctions do not themselves establish coherence in text, they help to identify the inferences writers imply between and among sentences. In other words, technical writers can make difficult concepts easier to understand by using conjunctions appropriately. Likewise, writers can mislead readers with inappropriate conjunctions. Similar to the linguists' classification of conjunctives based on semantic relations among clauses, Tichy (1966) suggests a guide that parallels this classification:

Table 6.1 Comparison of Semantic Relations to Tichy's Categories

Semantic Relation	Tichy's Categories
Additive *(AND)*	Addition: and, again, also, besides, equally important, finally, first, moreover
	Exemplification, repetition, summary, intensification: as has been stated, as I have said, as well as, for example, for instance, in fact, obviously, of course, that is, to be sure, to sum up
Temporal *(THEN)*	Time: afterward, at length, after an hour (day, week, year), immediately, meanwhile
Adversative *(YET)*	Comparison/contrast: after all, but, however, in comparison, in like manner, on the other hand, on the contrary, whereas
Causal *(SO)*	Result: accordingly, consequently, hence, therefore, thus, wherefore
	Purpose: for this purpose, to this end, toward this objective, with this goal
	Place: adjacent to, beyond, here, near, on the other side, opposite to

Lexical reiteration and collocation, reference established by pronouns and determiners, and semantic relations established by conjunctions are the most frequently used cohesive techniques in technical prose. In example 15, reiterative cohesion takes the form of repeated lexical items (e.g., *hyperactivity–hyperactive*), synonyms or near synonyms (e.g., *hyperactivity–hyperkinesis*), and general terms (e.g., *condition* for *hyperactivity*). Linguists call terms that belong to the same area of knowledge or the same topic (e.g., *hyperactivity, personality, anxiety, hyperkinetic*) examples of lexical collocation; however, technical writers know these are inevitabilities in technical writing. Whenever they introduce a main topic, technical writers develop that topic by creating cohesive "chains": the main topic follows a dominant chain of reiteration and collocation; secondary topics are linked to the main topic by similar chains. In (15) the main topic, hyperactivity, gives way to the specific secondary topic, assessing hyperactive behavior. Notice that the passage uses few conjunctions (see Table 6.1) but instead exploits the cohesive potential of pronominal/determiner reference (ref), reiteration (LR), and collocation (LC):

15. Descriptions of hyperactivity are generally given in behavioral terms, such

as motor activity, attention span, frustration toleration, excitability, impulse

                                                    (ref)   (LR)
control, irritability, restlessness, and aggressiveness. Although these behaviors
   (LC)   (ref)   (LR)
are measurable, they may not adequately reflect the kind of problems that
     (LC)   (LC)   (LR)   (LC)
different children have. Objective measures of attention span have been devel-
                      (LC)   (LC)
oped and have been used in some studies to help make the diagnosis, usu-
     (LC)   (LC)   (LC)   (LR)   (LC)
ally during tasks requiring continuous performance by the child. Standard
   (LC)   (conj)   (LC)   (LR)   (LC)   (LC)
questionnaires have also been designed to be used by parents and teachers as
(ref)   (LC)   (LR)   (LC)   (LR)   (LR)
they make observations at different times; such questionnaires provide useful
(LC)   (LC)
evaluation information. (Cited in Houp & Pearsall, 1984, p. 170)

Words alone do not establish cohesion; at least three major syntactic patterns assist in managing information flow and creating cohesion. Main and subordinate clause constructions, parallel constructions, and active and passive voice constructions position information in sentences so the writer can, respectively, emphasize some points and deemphasize others, focus readers' attention on a topic (Faigley & Witte, 1983), or move information to positions that are maximally cohesive.

Complex sentences (i.e., sentences with main and subordinate clauses) in functional prose are easier to understand when the main clause comes before the subordinate clause. For example, in (16), a description of a sequenced operation, the main clause *(turn off the power on both units)* emphasizes a necessary step in the sequence whereas the subordinate clause *(before you set up the connections)* emphasizes the temporal organization of the sequence:

16. Be sure that you have thoroughly read the instructions for the amplifier you intend to use with the V-4RX. Turn off the power on both units before

you set up the connections. (*Owner's Manual for TEAC V-4RX Cassette Deck,* 1983)

Parallel constructions give equal emphasis to a series of related ideas, make longer sentences easier to follow, and connect sentences to one another across a passage. Essentially, parallel structure is repetition with a syntactic twist: instead of repeating words, the writer repeats grammatical structures (e.g., prepositional phrases, noun phrases, verb phrases, or whole clauses) in a logical sequence (e.g., spatial, chronological, climatic, emphatic). Information in each repetition changes, but the grammatical form of the repetition stays constant. This paragraph began with a parallel series of verb phrases *(give equal emphasis, make longer sentences easier to follow,* and *connect sentences to one another across a passage)* that imply three equally important functions of parallelism: emphasis, readability, and cohesiveness.

Parallel structures enhance cohesiveness when used at the beginning of sentences or paragraphs. In (17) the first sentence begins with a prepositional phrase that announces a general class *(in several animal species).* The subsequent sentences repeat this grammatical form but restrict the class to one species per sentence. The subject of the article, mice, appears in the last position in the paragraph, typically a position for emphasized information.

> 17. *In several animal species* certain serum proteins appear to be different in males and females; for example, the relative concentration of albumin in rats, as measured by moving boundary and zone electrophoresis, was found to be higher in females than in males. *In cattle,* males possessed less a-globulin glycoprotein and more b-globulin and v-globulin glycoproteins than females. *In toads,* the separation of some of the serum components by starch gel electrophoresis has been reported to be different in the two sexes. *In mice,* the concentration agglutinating antibody to chicken and sheep heteroantigens, and to hyman erythrocytes, was found to be higher in females; and, in addition a protein fraction has recently been described as missing in male mouse serum (Cal-A strain) analyzed by a starch gel electrophoresis. (Cited in Tichy, 1966, p. 276)

Whether to use passive voice sentences is a debate of particular interest to technical writers. Grammar handbooks say avoid the passive, but functional prose abounds with it (e.g., example 15). Researchers, investigating readers' recognition of information in various sentence types, found that readers were consistently more accurate in verifying information in active voice constructions (e.g., 18a, c, and e) than in passive voice constructions (e.g., 18b, d, or f) (Clark & Haviland, 1977; Hornby, 1974).

> 18a. Active Voice:
> Space cities would greatly enhance the survival potential for *humans.* (*Broca's Brain*)

18b. Passive Voice:
The survival potential for humans would be greatly enhanced by *space cities*.

18c. Cleft/Active Voice:
It is *space cities* that would greatly enhance the survival potential for humans.

18d. Clef/Passive Voice:
It is space cities that the survival potential for humans would be greatly *enhanced by*.

18e. Pseudocleft/Active Voice:
The idea that would greatly enhance the survival potential for humans is *space cities*.

18f. Pseudocleft/Passive Voice:
The idea that the survival potential for humans would be greatly enhanced by is *space cities*.

Sentences (18a) to (18f) differ in emphasis and foci because of where information is placed in each sentence and how strongly the sentences mark certain information as new. In (18a), for example, the stress falls on the *survival potential;* this idea appears in the end of the sentence, a position typically filled by new information. However, (18b) switches the focus strongly to *space cities*. Information is more strongly marked when it is in a sentence-final position or when it appears in a *by* phrase (Clark & Haviland, 1977); in (18b) *space cities* fulfills both of these criteria. Because it begins with a nonreferential subject, (18c) also places focal stress on the *space cities* and puts secondary stress on *survival potential,* but (18d) emphasizes *enhancing,* while (18e) and (18f), though wordy, put focal stress back on *space cities*.

Using passive constructions is less a matter of linguistic dexterity than of recognizing what special functions passives can have. Passives can hide an unimportant or unknown agent or actor; they allow a more slowly paced, deliberate presentation of information; they promote stylistic variety; and they may function as weak substitutes for the imperative (Tichy, 1966). Most importantly, passive voice constructions allow writers to shift the object of a sentence to an initial position either to give it focus as old information or to use it as a connector to the new information presented at the end of the previous sentence.

Revising words, phrases, and sentences is not enough; paragraphs also need careful structuring. Contrary to popular advice, paragraphs are not arbitrary designations. Their purpose in technical writing is to direct coherence, the flow of ideas, by dividing a writer's thoughts into visual units. They are signals of substantive progression.

Effective paragraphs are rhetorical units designed for readers' convenience. They allow readers to comprehend information at two levels: the discrete units (i.e., the individual paragraphs) and the relationship among the units. Structuring paragraphs is like structuring sentences: writers use the syntactic devices of cohesion such as information distribution, reference, lexical cohesion, and syntactic arrangement (Meade & Geiger, 1970; Meyers, 1984). Readers typically use the first part of a paragraph to orient themselves and to construct a tentative framework for the coming information. From this early information readers make a preliminary assessment of the paragraph's topic, connect this topic to their world knowledge (through frames, schema, plans, and scripts) and to the text-presented knowledge (through previously textual information), and assume a development for the paragraph.

Given this dynamic, a topic sentence should begin the paragraph. Moving the topic sentence to other positions in a paragraph, as in the following variation of (3), can confuse readers:

> 3. An electron in an atom has a principal quantum number n (somewhat similar to the n in the Bohr theory), a second one 1 that restricts the size of the angular momentum of the electron's orbit around the nucleus and a third one that restricts the orientation of that angular momentum. New numbers popped out of the equation in orderly fashion.

Effective writers know the key to effective paragraphing: they abandon the concepts of paragraphs while writing and, instead, first compose their documents with their readers' needs for information in mind. Then they paragraph their material to guide the readers' attention. For example, the information in (11) falls naturally into two paragraphs, first because of the chronological organization and second because of the associated information (i.e., *instruments—narrow-angle camera—wide-angle camera*). The information in (19) also falls into two paragraphs not only because of its organization but also because its writer used questions to lead the readers into the topic:

> 19. How is it that a porpoise, a frog, a human being or a fruit fly achieves its characteristic appearance? What genetic mechanisms guide the development of a fertilized ovum so that it differentiates into the internal organs and physical structures that give an individual member of a species form and function?
>
> Some significant answers have begun to emerge from the studies of the DNA of the fruit fly *Drosophila melanogaster*. Two clusters of genes figure importantly in the work. One is the bithorax complex, first identified by Edward B. Lewis of the California Institute of Technology. This complex controls the development of the thoracic and abdominal segments. Another cluster, the Antenapedia complex, identified by Thomas Kaufman and his colleagues at Indiana University, shares in the control of thorax development and promotes the development of the head. (*Scientific American*, Sept. 1984, p. 82d)

The two criteria for effective paragraphing, composing documents for readers' needs and using paragraphs as visual guides for readers, bring this discussion full circle. What began as recommendations for revisions at a text level ends, logically, at that same point. By organizing information coherently at a text level, revising how that information is distributed in the document's sentences, and formatting the information in paragraphing, writers can create and signal the coherent structure of their documents, making their work what it should be: functional and informative.

References

Bain, A. (1877). *English composition and rhetoric.* New York: American Book Company.

de Beaugrande, R., & Dressler, W. (1981). *Introduction to text linguistics.* New York: Longman.

Clark, H. H., & Haviland, S. E. (1977). Comprehension and the given-new contract. In R. O. Freedle (Ed.) *Discourse production and comprehension* (pp. 1–40). Norwood, NJ: Ablex Press.

Faigley, L., & Witte, S. P. (1983). Topical focus in technical writing. In P. V. Anderson et al. (Eds.), *New directions in technical and scientific communication: Research, theory, practice* (pp. 59–68). Farmingdale, NY: Baywood.

Flower, L. (1981). *Problem-solving strategies for writing.* New York: Harcourt Brace Jovanovich.

Halliday, M. A. K., & Hasan, R. (1976). *Cohesion in English.* New York: Longman.

Hornby, P. A. (1974). Surface structure and presupposition. *Journal of Verbal Learning and Verbal Behavior, 13,* 530–538.

Houp, K. W., & Pearsall, T. E. (1984). *Reporting technical information* (4th ed.). New York: Macmillan.

Jordan, M. P. (1974a). Besides moreover however and but conjunctions—Order out of confusion. *Journal of Technical Writing and Communication, 4*(2), 133–146.

Jordan, M. P. (1974b). Next then although more too—Conjunctions in action. *Journal of Technical Writing and Communication, 4*(3), 171–183.

Jordan, M. P. (1979). Some background theory of clause relations. Mimeo. Kingston, Canada: Queen's University.

Jordan, M. P. (1982). The thread of continuity in functional writing. *Journal of Business Communication, 19*(4), 5–22.

Jordan, M. P. (1984). *Fundamentals of technical description.* New York: Krieger.

Larson, R. L. (1967). Sentences in action: A technique for analyzing paragraphs. *College Composition and Communication, 18*(1), 16–22.

Markels, R. B. (1984). *A new perspective on cohesion in expository paragraphs.* Carbondale: Southern Illinois University Press.

Meade, R. A., & Geiger, E. W. (1970). Paragraph development in the modern age of rhetoric. *English Journal, 59*, 219–226.

Meyers, D. (1984). *Understanding language.* New York: Boynton/Cook Publishers.

Schank, R. C., & Abelson, R. P. (1977). *Scripts, plans, goals, and understanding: An inquiry into human knowledge.* Hillsdale, NJ: Erlbaum.

Tichy, H. J. (1966). *Effective writing for engineers, managers, scientists.* New York: Wiley.

Winter, E. O. (1977). A clause-relational approach to English texts: A study of some predicative lexical items in written discourse. *Instructional Science, 6,* 1–92.

Winterowd, W. R. (1970). The grammar of coherence. *College English, 31,* 828–835.

7

How Can Text and Graphics Be Integrated Effectively?

M. JIMMIE KILLINGSWORTH and MICHAEL GILBERTSON

Although readers expect functional documents to display information visually, technical writers plan how visual displays can be effectively integrated with text so that the text and the displays are both persuasive and stylistically objective. Writers must decide what type of visual display the data should have, how many visual displays are necessary, and where the displays should appear in a document. Technical writers understand that poorly designed or inappropriately placed graphics can confuse readers and prevent them from making accurate judgments based on the information in a document. Further complicating the issue is the multiple audiences for technical documents. Different audiences using a document for different purposes will not need the same graphics.

Professors M. Jimmie Killingsworth and Michael Gilbertson argue that designing visual displays and placing them in a functional document are rhetorical decisions. Selecting and placing visual displays are similar to the textual decisions writers make when they create a persuasive argument or a logically coherent report or revise to insure an objective style. For example, writers of reports are concerned with quantitative and qualitative information; thus, the graphics in reports must efficiently arrange data and present it with impact. In technical manuals, writers emphasize how to perform a specific task; thus, the graphics in manuals and instructions must simplify information, focusing the readers' attention on specifics. Writers of proposals establish credibility in order to sell ideas; thus, the graphics in proposals or promotional information must stimulate interest while counteracting readers' potential uncertainties.

Because visual displays should be fully integrated parts of a document, the authors recommend that writers incorporate graphics in the writing process. When composing, writers should sketchily illustrate a document as they write it. Writers must collaborate with artists as they plan the document so that their words interact with the artists' contributions. Writers must remember the purpose of visual representations is to add impact.

The Problem and Approach

A major problem in technical communication involves how graphics should relate to the written text. Obviously visual and verbal elements should be effectively integrated. But to what degree graphics complement, supplement, replace, enhance, emphasize, summarize, decorate, overwhelm, or distract from the written text may baffle the student, teacher, or new practitioner of technical writing.

Finding a solution to the problem is, first of all, a matter of rhetoric (see Chapters 3 and 4 for further discussion). The technical writer must find an approach to the audience that will at once accommodate verbal and visual material. The rhetoric, that is, must be comprehensive. In addition, the solution involves practical concerns. Proper integration of graphics and writing requires planning and thinking visually as well as verbally in all stages of document development.

In this chapter, therefore, our approach is twofold. Part 1 is audience oriented—a discussion of how authors can come to view graphics as an integral part of their written appeal to the readers and users of technical reports, manuals, and proposals. Part 2 is author oriented—a discussion of the process by which writers develop the visual elements along with the written portions of their documents. In Part 1 we lay the theoretical foundation for our consideration of the practical concerns of Part 2.

Part 1: A Rhetoric of Relevancy for Verbal and Visual Communication

Relevant Research

Technical communication is action oriented. Its context is business, government, and industry (and the university insofar as it feeds into and draws support from these other mainline American institutions). The primary purpose of all technical documents is to get things done.

Given this purposefulness, the key concept in a rhetoric for technical writing is *relevance*. A technical document should be relevant to a specific *audience* and a specific *task* or project. Both the author and the audience are somehow involved in this task. The author may be requesting funding for a project, reporting on its outcome, or helping the audience enact some

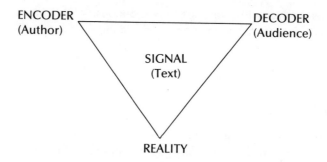

Figure 7.1a The Communication Triangle

portion of it. Figure 7.1 presents Kinneavy's (1971, pp. 17–40) "communication triangle" and a revision of it that represents a practical model of technical communication. In this latter scheme, the document serves as the link between author and audience; all three of these interrelated elements are task directed. That is, relevant information helps or encourages the audience to accomplish the task, or it explains the outcome or projected outcome. Irrelevant information slows, inhibits, or in other ways discourages action, and it obscures explanations.

Tufte (1983, p. 182) shows that the purpose of good graphics is either to communicate and illustrate a settled finding or to explore the possibilities of a data set; in either case, the aim of visual display is to *reveal data*. This is the visual correlative of the informative aim of technical writing: *to disclose information*. Some may question whether rhetoric, in its classical and academic sense, has a place in the disclosure of information and the revelation of data (see Cox & Roland, 1973, pp. 140–142). But unlike history and journalism, technical writing discloses only that information relevant to a given task. And once relevance is accepted as a guiding principle, direct and modified applications of classical rhetoric and modern discourse theory become obvious. Making information relevant quite clearly

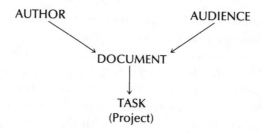

Figure 7.1b The Revised Communication Model for Technical Communication

involves *discovery* and *selection*. Together, as invention, they form the first of the major operations of rhetoric. The second two, *arrangement* and *presentation* (style), also come into play with the summoning of a specific audience. Highly technical displays of data—a high-density line chart, for example—may be arranged in appendixes for the benefit of a secondary audience if the primary audience of a report is a company executive interested mainly in the overall management of a project. Such a chart would be an effective—and thus relevant—presentation for an audience of engineers or accountants but could frustrate readers uninterested in mathematical detail.

Classical and modern rhetoric helps in deciding such points. To see clearly the applications of such theory, we divide technical communication into three modes, each with its specific forms and varying aims and each drawing on a somewhat different portion of rhetorical theory. We call the three modes the reportorial, the operational, and the promotional. A summary of this classification appears in Table 7.1.

The three modes relate the author's intention directly to a task and audience. They distinguish between the three temporal contexts in which actions occur—past, present, and future. And each of them corresponds, largely on the basis of this temporal distinction, to one of the three types of speeches discussed in Aristotle's *Rhetoric*. Because of the varying intentions and temporal contexts, each mode requires a different approach to the audience and each has a different set of rhetorical concerns. In the part of our chapter dealing with the three modes of technical communication, then, we are concerned with the "audience side" or the psychology of our communication model, as shown in Figure 7.2.

Table 7.1 The Modes of Technical Communication

| Mode | Major Form in Technical Writing | Aristotle's Rhetorical Type | Time Conveyed | Author's Intention in Presenting Task to Audience | |
				Primary	Secondary
Repertorial	Report	The forensic speech	Past	To confirm and explain	To explore and recommend
Operational	Manual	The epideictic speech	Present	To help accomplish and find information (tutorial and reference)	To explain and encourage
Promotional	Proposal	The deliberative speech	Future	To project, explain, and persuade	To explore possibilities

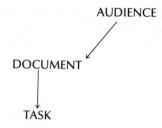

Figure 7.2 The "Audience Side" of the Model for Technical Communication

The Reportorial Mode

Authors of technical reports deal with past events: a report confirms that a project occurred, gives the results of the project, and usually explains the results. Though the report deals with the past, the information must be relevant to the reader's present needs. As Barzun and Graff (1970, p. 5) have noted, the main distinction between the historian's interest and that of the technical report writer is "the former seeks to know the past; the latter is concerned with the present, generally with a view to plotting the future." Thus, writers face a gap between what may be called the *actual* and the *rhetorical* time of the events they report.

Aristotle recognized the same problem in forensic discourse, which also deals with past events as they relate to present judgments. The forensic speaker must demonstrate his or her competence (ethos) through logical and exact exposition (logos). This obviously involves careful selection and logical arrangement of facts. But the concern for the audience and the task also suggests that the author may convey a sense of immediacy, if not urgency, in the report. Thus, the style must incorporate active, vivid, and direct language.

The report therefore requires presentation that is at once "quantitative" and "qualitative." Technical reporters must write in a style that is exact (quantitative) but also vivid (qualitative). And they must use the full range of techniques for visual presentation to meet the same ends. Graphics satisfy the quantitative demands by neatly and efficiently arranging data and meet qualitative demands by providing impact.

Because of their quantitative power, graphics are certainly superior to verbal statements and can be gainfully used as substitutes for words in many situations, especially where exact presentation is required. Part of the power of good graphic displays lies in their efficiency. They allow audiences to take in information *at a glance*. Lefferts (1981, p. 5) claims that "less time is required . . . to comprehend information when graphic methods are employed"—a point that the literature on visual psychology seems to support: "A pictorial image presents itself whole, in simultaneity" (Arnheim, 1969, p. 249). A correlative advantage is that graphics are "commanding" and have even come to be expected in technical documents (Murgio, 1969, p. 34). The marshalling of numerical data contributes to the ethos of the document by lending an air of authority, by conveying "a sense of surety and exactness" (Lefferts, 1981, p. 31)—hence, the impact of graphics.

These very advantages, however, suggest the need for caution, for graphics can be overused and misused. As Murgio (1969, pp. 21–22) points out, charting can be carried to an extreme in material that is "too simple" ("sales went down"), "too familiar," or even "too specific": "In situations where the exactness of double-entry bookkeeping is required, charting is usually uncalled for." Visual irrelevancies, such as the gratuitous "little man" cartoon and other forms of "chartjunk" (Tufte, 1983, p. 121), may add impact at the cost of distracting from informative points or trivializing the report. Moreover, though some authors claim that a report seldom contains too many graphics (Stratton, 1984, pp. 115–116), others warn that the high density of the graphic form occasionally leads to "saturation" of the report's senses (Magnan, 1971, p. 59; Rockett, 1959, p. 59; Tufte, 1983, p. 26ff).

Besides overuse, many critics have commented on the misrepresentation of statistics in graphic form (Huff, 1954; Murgio, 1969, p. 39; Tufte, 1983, p. 13). Truncating charts, distorting the scale, and other such misrepresentations measure the limits of persuasive rhetoric in technical reporting. Accurate and honest communication is always the goal; often it is a legal requirement as well. Technical communicators should familiarize themselves with common graphical abuses to avoid inadvertent misrepresentation.

Furthermore, the report writer needs to recognize that graphics not only represent data but also interpret them. It is a mistake to think that the business of interpretation is left entirely to the verbal text. This is especially a problem because in most technical writing the author aims for objectivity. To leave a chart without verbal interpretation does not necessarily free the report from subjectivity; it merely leaves the audience to glean what significance is available from the chart by itself.

Graphics allow interpretation by providing through simplification a means for classification based on comparisons and other relationships. Bertin (1981, p. 7) explains that "simplification is no more than regrouping similar things. The eye simplifies by correcting the irregularities it notices in . . . initial disorder." Thus, "visual perception is spacial perception [that] allows anyone to use a new system of classing: the simultaneous consideration of different elements."

The information contained in a report is not merely individual items of

Table 7.2 An Example: Enrollment in Writing Courses

	1983			1984			1985		
	Spring	Summer	Fall	Spring	Summer	Fall	Spring	Summer	Fall
Total college enrollment	1250	692	1287	1221	632	1237	1195	567	1220
Freshman composition	120	60	150	115	50	142	117	52	148
Advanced composition	8	10	15	6	10	21	12	7	14
Advanced report writing	12	6	20	14	7	16	13	5	18
Editing	11	5	14	16	8	12	21	6	13
Manual writing	7	7	12	13	5	15	17	7	11

data; it is, as Bertin (1981, pp. 12–13) notes, the relationship possible among elements, subsets, or sets. These divisions constitute three levels that must be retained in the graphic if it is to have full interpretive power:

1. *The elementary level.* Every entry along the x and y axes represents a single element as it is placed in relation to other single elements. At this level the example (Table 7.2) shows, for instance, that 14 students were enrolled in Advanced Report Writing in the spring semester of 1984. Simple as this is, imagine the amount of verbal text that would be required to state all of the elementary relationships contained in the table.
2. *The intermediate level.* At this level readers of the report, guided by the author's verbal text or through their own exploration of the data, can explore the relationship within subsets such as summer, spring, or fall, and answer questions like, "What are the most popular courses in summer?"
3. *The overall information level.* Decisions occur at this level since it answers questions about overall trends in the relationship of x and y (time and enrollment), such as "When do students tend to enroll most heavily in advanced courses?"

Bertin (1981) makes a strong case for the trilevel interpretive power of graphics. He concludes that "a construction which does not enable us to define groupings in x and y does not reach the overall information level of the entire set. This is the mark of inefficiency" (Bertin, 1981, p. 13). The implication is that efficiency in decision making depends on the kinds of groupings that graphics provide and that words can, at best, weakly imitate.

We must add, however, that not only efficiency but also relevancy is the aim of report writers. They display data and sometimes make or influence decisions. But they are often called upon to explain the data. In this case the author must guide the reader by posing relevant questions. Verbal text remains the "voice" and the guiding hand of the author. Such guidance becomes increasingly necessary as one moves to increasingly higher levels of information. A possible exception is the occasion when the report involves the exploration of data, as in reports of investigations that are still inconclusive at the time of writing. Even here, however, readers are rarely left to consider a table and find what they may.

The exploratory aim of some reports suggests another means of presenting visual material—not tables and graphs, which almost always require the verbal intrusion of the author, but functional diagrams represent abstract relationships that become clearer in visual than in verbal form and thus may be used to supplement the verbal text. Several authors contend that graphics do not easily accommodate abstract ideas, which are better embodied in words (e.g., Magnan, 1971, pp. 739–740). In fact, visuals work quite well except in cases of high-level abstractions (Murgio, 1969, p. 19) or statements in which several levels of abstraction are applied to a single entity, such as "Lions are cats" (Arnheim, 1969, p. 239). Arnheim distinguishes between intellectual thinking and intuitive thinking. The latter is most useful in exploratory mental activity like that required in theoretical science, and, as it turns out, it is effectively spurred by visual presentations that allow the entirety of a problem to be grasped at once. Language is often too analytical and tends to isolate the parts of a problem from one another.

Hence the concept of the "model" arises. Modeling many times involves new ways of looking at the physical world, as in the case of Newton's "perceptual transformation" in learning to think of weight as an effect of gravity. Modeling also allows a person to "shape" thought by granting abstractions a kind of spatial existence. According to Arnheim (1969, p. 116), the nonmimetic images involved in such thinking find their communicative prototype in "those diagrammatic scribbles drawn on the blackboard by teachers and lecturers in order to describe constellations of one kind or another—physical or social, psychological or purely logical." By inserting between abstract words, lines, or arrows representing force, influence, and direction, an author helps the audience to see as a whole a process or set of relationships that would be atomized by placing them in sentences and paragraphs. Figures 7.1 and 7.2, depicting our model of technical communication, are examples of such diagrams.

These figures retain the power of presentational graphic charts to condense information and present it in its entirety, and they add to the technical writer's methodology a visual technique for dealing with abstractions. They are closely related to other functional diagrams like the flowchart, which depicts processes, and the line drawing, which describes objects and mechanisms. We will see the significance of these graphics as we turn to our discussion of the operational mode.

The Operational Mode

Technical manuals and other such instructional documents represent in many ways the purest application of our communication model. They deal directly with a task in which the reader (user) is immediately (presently) engaged.

But too often manuals are product oriented rather than task oriented. As such they neglect the reader's needs and focus instead on the attributes and capabilities of the product or system. Their authors confuse the reportorial with the operational mode. In a computer manual for a software package, for example, the structure of the system may dominate the organization of the document. Sections on the file manager, linker, and loader may replace task-oriented sections that present the operations of the program. Maynard (1982) suggests that the manual's chapters should be divided on the basis of each operation's complexity and frequency of use. The first chapter should contain the simplest, most frequently used information, and in each chapter thereafter the operations should increase in complexity and decrease in frequency of use. In this manner the reader's relationship to the task becomes the foremost consideration of the manual writer.

Emphasis on the product rather than the task may also render the choice of visual material ineffective. Quite obviously a photograph, for instance, may project the most realistic and thorough visual image of a product. But photography, even with airbrushing and other retouching techniques, tends to be unselective. As Titen (1980, p. 115) notes, "where function rather than form is being demonstrated, details of physical appearance become superfluous." The reader encountering a flood of detail may be distracted from details vital to the performance of the task. Thus, the photograph may give too much or too little information (Turnbull & Baird, 1975, p. 170). Its use should be limited to general introductions and orientations to the product. Over-the-shoulder shots of a user at a computer keyboard, for example, may align the manual with the user's viewpoint.

The author of tutorial manuals should opt most often for line drawings that focus the readers' attention on the pertinent details involved in each task. In user manuals for computers, figures accompanying operating instructions may repeat the same generalized drawing of a terminal or keyboard with different keys or buttons shaded or colored for separate operations. The only necessary details are those that allow users to find quickly the parts of the machine that must be used in the task at hand. Authors of maintenance and repair manuals should use schematics and exploded views instead of photographs. In addition to the advantage of emphasis and focus, they add flexibility to the presentation.

Line drawings and other functional diagrams should be preferred in operational documents for psychological as well as practical reasons. Just as Aristotle lays great stress on the development of ethos in speeches that deal with present realities *(epideictic),* many commentators on manual writing have suggested the need for developing a special illusion of personality in the document's tone. Skees (1982, p. 181) advises, "Look upon the user manual as the document which substitutes for your own physical presence, providing everything in the way of guidance and assistance that you, yourself, would provide to the user of the system if you were there." The line drawing has the advantage over the photograph in enacting this rhetoric because, as Arnheim (1969) notes, line drawings derive from descriptive gestures. They are, therefore, the ideal accompaniment to conversational style in developing an integrated verbal/visual text designed to project a tone of helpful concern. Cartoons may also contribute to the air of relaxed informality.

Turnbull and Baird (1975) have noted that designers of layouts and artwork for newspapers, magazines, and ads must develop design principles for attracting and arousing attention, since readers' attention is often low and their interest span short—very unlike the readers of books whose attention and interest tend to be high. A parallel distinction exists in technical communication.

In all modes it is possible that the reader will be hurried or reluctant, but in the operational mode it is almost certain. Although clear headings and other information locators and readability aids are needed in all technical documents, such devices are even more crucial in manual design. As opposed to manual readers, report readers tend to have a higher level of interest since they are usually decision makers. And though proposals could certainly benefit from all the techniques of arousal and influence pioneered by the developers of mass media and advertising theory, they must not arouse suspicion in the audience; they must build a case on the merit of content and must therefore avoid the "slick" appearance of a "snow job." In proposals to federal agencies, regulations expressly forbid "brochuremanship" (Oslund, 1962, p. 28).

In addition to making the manual more pleasant and easier to use, such devices of format as headings, white space, color, and highlighting may be used to help the reader comprehend the manual's information. When instructions are complex, tables, lists, and flowcharts are more effective than prose paragraphs. In describing processes, for example, flowcharts can accommodate recursion and branching much more clearly than can sentences, which tend to be highly linear in structure. Drawing on research in cognitive psychology, Rude (1985, P. RET 36) has found that "format affects comprehension in essentially two ways: 1) it helps a reader sort information and perceive its structure; [and] 2) it gives learners control over learning by facilitating selective learning." Good layouts can thus enhance both the tutorial and the retrieval functions of the manual.

A final note on manual design: some technical communicators may be tempted to bypass the problem of integrating text and graphics by resorting exclusively to "wordless instructions" or pictorial manuals. Except for very simple tasks, such as changing a typewriter ribbon, such presentations are not very effective. Visual psychologists have determined that verbal instructions definitely affect eye movement in viewing pictures and can therefore profoundly assist the reader in receiving and sorting information (Tufte, 1983). Moreover, early results in research on wordless instructions suggest that the pace is so fast that comprehension is prohibited or inhibited (Scofield, 1977).

The Promotional Mode

Promotional writing deals with the future. It aims to persuade the audience to adopt a certain course of action. Sales literature, for example, encourages the reader to buy a product. A proposal outlines a plan for action that requires the reader's funding or permission. All technical documents in the promotional mode must, though they deal with the future, address the present needs of the audience. Like the report, therefore, they must face a gap in actual and rhetorical time.

The proposal writer must establish credibility, which is a form of relevance, since a credible document (like a credible fiction) treats characters, problems, ideas, and facts that are relevant in some way (logically, empirically, or psychologically) to the reader's reality. In proposal writing such materials have, or appear to have, the tangibility of the past and present, a definiteness that offsets the uncertainty of the future. The successful proposal accomplishes three goals (Holtz, 1979, Chapter 14):

1. It creates interest and desire in the audience.
2. It dispels the audience's fears about the uncertain future.
3. It demonstrates clear logic.

Interest and desire are partly matters of visual presentation, though prohibitions on "brochuremanship" may severely limit the technical author and graphic artist. Still, simple devices in the layout and format may attract attention as well as improving readability. Holtz and Schmidt (1981) recommend the use of headlines and "blurbs" or glosses containing words with impact value and avoiding standard tags like "introduction" and "general background" (unless these are specified in the Request for Proposal). If color and fancy typography are prohibited, photography should be considered. Turnbull and Baird (1975, pp. 98–100) note that mass media experts consider the photograph to be the "best means for attracting a reader's eye to a page. . . . The emotions or reactions that are aroused as we view life about us can be aroused and catered to by photographs better than any other means."

In addition to stimulating interest, photographs help to achieve that second goal of promotional documents—counteracting uncertainties about the future by building a sense of a specific present reality. More often than not, however, proposals deal with products and tasks that are still in the design stage. Since it is impossible to photograph products that do not yet exist, realistic mock-ups or prototypes may be developed and photographed. Or other means must be found "to make the abstract, concrete; the unfamiliar, real; the vague, definite" (Beatts, 1959, p. 7). In engineering projects, schematics have the necessary simulating power: "They show on paper the circuit essentials, as if the circuit were [itself] available" (Beatts, 1959, p. 10). Time charts and organization charts likewise add concreteness to management plans (Lefferts, 1981, pp. 110–111, 133). And, for nontechnical audiences, a similar vividness arises from an "application illustration," a drawing or painting whose purpose is to demonstrate "the actual use of a product or equipment under conditions in which it was designed to operate, rather than to depict the exact details of the mechanism"; in a military proposal, for example, a tank may be shown plowing through dense tropical terrain (Magnan, 1971, p. 760). Illustration may thus be used to "dramatize, as well as to communicate more effectively" (Holtz & Schmidt, 1981, p. 246). By enhancing the vividness of the evaluator's mental picture of the future project, illustration grounds the proposal in an illusion of present reality and thereby increases its credibility and effectiveness.

Careful logic is necessary in successful proposals not only because it justifies the proposer's plan but because it also helps allay the fears and suspicions the evaluator may have about the future. Since it is a common system of thinking among educated people, logic makes them comfortable by leading them into a frame of mind they have entered many times before. In a discussion of future-oriented deliberative discourse, Aristotle recommends using examples to build an empirical base of factual and logical (enthymematic) information (Freese, 1982). One must use logic and imagination to extend the present into the future, since facts about the future do not exist. Charts that show trends or changes over time are particularly effective in proposals because they take factual information from the past and present and project it logically into the future (see Lefferts, 1981, pp. 81–82; MacGregor, 1978, p. 106). Tables can often be converted to line graphs that dramatize trends more effectively.

As Bicking notes, "The higher one looks in administrative levels of business, the more one finds that decisions are based on data that are analyzed statistically and presented in tabular or graphic form" (quoted in Enrick, 1972, p. 166). It is the obvious work of proposals to stimulate this kind of decision, so the proposal writer must become familiar with visual methods for displaying and evaluating statistical information.

Part 2: A Composing Process That Helps to Integrate Text and Graphics

Recommendations

Our discussion so far has been oriented toward the audience and the product (document) of technical writing. But the most interesting innovations in the literature devoted exclusively to technical communications focus on the process by which documents are created; it deals, that is, with the author side of our communication model (Figure 7.3).

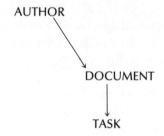

Figure 7.3 The "Author Side" of the Model for Technical Communication

This trend conforms nicely to a principle that has recently emerged as a dominant force in composition theory. Researchers in the teaching of writing have found that best results are produced when teachers concentrate on the *writing process* and deemphasize the final product. In an essay summarizing this movement, Hairston (1982, p. 84) writes, "We have to try to understand what goes on during the internal act of writing and we have to intervene if we want to affect the outcome. We have to . . . examine the intangible process, rather than . . . evaluate [only] the tangible product." The commentators on technical writing have in recent years made progress in making this "intangible" process somewhat less abstract and more accessible. Visual communication has greatly aided in this effort.

Despite the new emphasis on the author and the process of composing documents, audience psychology should never be neglected. Instead, "audience and intention should affect every stage of the creative process" (Hairston, 1982, p. 84). With the rhetoric of relevance in the background, then, let us consider the stages of planning, drafting, editing, and producing the verbal and visual texts.

One note of warning: In dividing the process into these parts, we must remember that in practice they are not neatly separated. The steps in the process are overlapping and recursive. Moreover, the last step, production, deeply influences the other steps. Certain kinds of graphic illustration must be eliminated during planning, for example, because production costs are prohibitive.

Planning the Document

Although many textbooks still suggest the priority of verbal expression, working technical writers have come to realize that illustrating a document after the text is written is a bad idea. "A visual aid must be made part of the text, not an ornament to it," write Mathes and Stevenson (1976, p. 173), but they add the stipulation that "a visual aid" (the very term implies the priority of the verbal text) "must be verbally interpreted." This is yet another variation of the standard textbook advice that "all visual aids . . . must be referred to in the text itself" (Weisman, 1980, p. 179). Clearly these suggestions are product oriented. They tend to lead to weak integration of text and graphics. *Technical Writing: Process and Product*, a recent textbook whose title denotes a familiarity with contemporary composition theory, takes a step in the right direction. The student is told, "As you write your initial draft, put in references to tables and figures wherever you think an illustration might be useful. Don't stop to prepare the illustration at that time, though" (Stratton, 1984, pp. 115–116).

In fact, proper integration of text and graphics begins well before the drafting stage. "Graphic ideation precedes graphic communication" (McKim, 1980, p. 111). As a project takes shape, the earliest notes, which become the seeds for the later document, should contain rough sketches of processes, mechanisms, and abstract relationships or models. A worthy prototype for such a notebook is that of Leonardo da Vinci. Leonardo was able to "move astutely back and forth between the verbal and visual to best convey ideas"; he thus "avoided the practice of modern authors who merely illustrate a text after having written it. . . . Leonardo thought visually or verbally according to the circumstances and subject" (Hughes, 1985, p. 30). The notebook should arrange data in rough tables and charts which, like the verbal text, can be edited later. Berlin (1981, p. 16) in a mild overstatement writes:

> Any problem can be conceived in the form of one double-entry table and thus can be transcribed in the form of a matrix. This is the most common solution, one which makes full use of the properties of visual perception.

Likewise flowcharts prove useful in early organizational thinking. Creating useable flowcharts is "a discipline primarily, rather than an art" (Holtz & Schmidt 1981, p. 247). As early as the "initial brainstorming session" in developing a proposal, a flowchart can be "a highly valuable tool for developing overall strategies and approaches": "it *forces* decision making" by lending an air of specificity to nebulous plans and obscure abstractions (Holtz & Schmidt, 1981, p. 247).

When the time comes to plan the organization of the document, visual thinking can play a key role once again. If an outline is used, its topics and subtopics provide natural guidelines for headings and matters of format. The outline also suggests possibilities for lists, flowcharts, and other devices of process and classification.

Presentational and functional graphics can even replace outlines and other organizational tools. In an early article on engineering reports, Rockett (1959, p. 61) recommends that writers simplify the presentation of documents by presenting as much content as possible in tables, charts, diagrams, drawings, and photographs. The verbal text should merely "supply the continuity." The same strategy forms the basis for the Graphics-Oriented (GO) method of proposal preparation (Green, 1985, p. VC30):

> A GO proposal's . . . illustrations, photographs, tables, schedules, logic flows and activities networks . . . communicate [the] company's approach to solving a customer's problem. In practice, these graphics are formatted and completed first, prior to any text preparation. . . . The percentage of graphics to text is a large one. The text merely provides adhesive for holding these data elements together, and it reinforces themes and advantages by highlighting the data points of greatest significance.

This methodology finds theoretical support in psychoanalysis and visual psychology. The problem of relating the individual elements in a series of visual images was addressed early in the century by Freud. Dreams, obvious examples of purely visual thinking, must in the absence of verbal explanation be filled with elaborate symbolic imagery, the "dream work." It uses "visual puns" and other graphic imitations of language to tie together apparently unrelated scenes and images. As Arnheim (1969) observes, the relation of the dream-work to the dream suggests an obvious use of language in daylight communication: it is more adept at providing logical transitions than are visual images. A purely pictorial text must remain a puzzle unless it depicts a simple chronological sequence of events, with one step following neatly after another. Such a simple process may in fact need no verbal explanation. It may be the occasion for a wordless manual, such as those employed by IBM to show how to load the correcting ribbon into a typewriter.

Drafting the Document

Tufte (1983, p. 180) registers a justified complaint about the process of document development and its often disastrous effect on the quality of the product: "Words and pictures are sometimes jurisdictional enemies, as artists feud with writers for sparse space. An unfortunate legacy of these craft-union differences is the artificial separation of words and pictures."

If the verbal text is to be properly integrated with the visual elements, authors and artists must cooperate at every stage of the compositional process. The same is true of anyone who contributes visual material to the document—scientists with their charts and diagrams, engineers with their schematics, graphic artists and editors with their designs for layout, and photographers:

> Working as an integrated team, the writer, designer, and photographer should meet prior to photography to discuss the subject in detail. If the team works in close association, the final results will reflect this, manifesting the three elements—unity, coherence, and emphasis—that are so essential to all successful picture stories. (Aumuller, 1971, p. 818)

Technical writers must know the capabilities of their fellow workers on the document team. They must also familiarize themselves with the possibilities offered by visual communication, even if they lack the skills necessary for producing finished work. There are many excellent texts that provide this information. (See especially Lefferts, 1981; Magnan, 1971; Murgio, 1969; Tufte, 1983; Turnbull & Baird, 1975; and Chapter 9.)

During the drafting of the document, visual techniques also provide a means of organizing the various contributors of verbal and visual material—a task that often falls to the technical communicator. Company experience often reveals effective formulas for developing the various components or modules of a document. The contributors may be handed a preformatted module in which to develop their particular portions of the presentation, whether verbal or visual. Space and size can thus be carefully maintained.

The modular approach to drafting documents may stifle creativity, however. One way of loosening the process up is to use a method pioneered by moviemakers—the storyboard. Proposal writers and managers have already become convinced of this technique's effectiveness (Englebret, 1972, pp. 115–118; Tracey, 1983, pp. 75–76); and it should work equally well with reports, especially those written incrementally as the project develops and with manuals that, like films, deal with dramatized action. Storyboarding involves placing the various verbal and visual segments of a document on a bulletin board as they are completed. The modules can be arranged and rearranged as the work evolves into a finished product. Thus, interaction between team members is improved. And this kind of presentation encourages maximum use of visual materials that can be easily and quickly comprehended and compared by fellow workers. It rapidly becomes clear when an illustration or portion of text is unsuited for the preestablished plan; then team members can decide whether the plan or the innovation is at fault.

In drafting their portions of the document, writers should remember the relationship of the verbal and visual texts. The visuals in reports require the writer to emphasize the relevant details and ask the appropriate questions. Manuals need instructional interaction between words and pictures. Proposals demand a vivid, specific style that helps the graphics give substance to vague projections into the future. Words are the most effective medium for providing historical and operational contexts and for narrating examples of relevant events.

In general, written text can be written and structured "to encourage reader visualization" (Mattingly & Whitesell, 1985, p. VC34). The verbal style must use strong, active verbs, for example, and vivid imagery (see Chapter 6). In addition, the "skeleton" of a technical document—its organization—must be brought to the surface by various cohesive devices including headings and subheadings (see Chapter 5).

Editing

Incremental drafting, the modular approach, and storyboarding allow editing to occur at all stages of document development. It can no longer be considered a last-minute clean-up operation. Moreover, it is not only the work of a specialist with the title of "editor." Just as the drafting process includes contributions from many members of the team, so should editing. For late revisions, however, a technical editor not involved in the early stages of the project may be more objective and more likely to make crucial changes. Both visual and verbal text should be edited with a clear eye for purpose and audience.

All the major forms of technical communication require visual presentations that promote *density, simplicity,* and *impact,* each of which is enhanced by a careful editorial hand. But as one shifts from mode to mode, emphasis tends to fall on one or another of these effects.

The reportorial mode favors density. During editing, charts and tables can often be combined, or a photograph may replace a line drawing to increase density and evidential value (Gross, 1983). Pie charts, because of their low density, should rarely be used in technical reports (Gross, 1983; Tufte, 1983); their effect is almost purely one of impact. The editor should also question matters of scale and labeling and should insist on eliminating unused space and superfluous wording within the graphic (Amsden, 1980). One should also realize that, as density increases, there is greater pressure on the verbal text to narrow and direct the audience's attention. Moreover, the editor must always remember that high density has a negative side: even the specialized reader can quickly become "saturated" or even confused by large amounts of information.

Simplification is implicit in the very construction of a graphic (Bertin, 1981), but manuals and other operational documents have a need for special simplicity. Line drawings, lists, and other devices that promote quick location and easy comprehension of information must be highly analytical; they must deal with one step at a time unless their function is merely to summarize or orient. The editor may need to divide illustrations. Labels and cross references may need clarification. Above all, the format should allow for easy reading and a clear correspondence between verbal and visual elements.

The visual parts of a promotional document are primarily responsible for impact, since even vivid language must be read before it can have any effect. The exhausted evaluator of a proposal or a casual peruser of sales literature may turn over pages of excellent prose only to stop at one striking photograph or a tastefully laid-out page whose verbal and visual elements are carefully balanced and contrasted. The objective editor is most valuable for this kind of document because distance is needed to determine impact.

In considering the written text, editors should aim for maximum visualization. If any of the verbal text can be converted to graphics, it should be (Amsden, 1982).

Production

Printing or duplicating the document is often out of the hands of the technical communicator. This is unfortunate since many times printing conventions designed for expediency and not rhetorical effectiveness may govern the possibilities of communication. Tufte (1983, p. 18), whose own company published his excellent text on graphics, suggests that communicators should do what they can to gain control of the final stages of document production. After all, why work hard to integrate graphics and text if a printer is going to disintegrate the unified text by refusing, for example, to print graphics on facing pages or by raising prices for slight modifications in conventional printing practices?

Communicators should thoughtfully write specifications for printing, carefully supervise duplication work, and insist on quality control. They should develop a clear understanding of the procedures and capabilities of typesetters and printers through formal college training, reading (in such works as Turnbull & Baird, 1975), and thorough observation on the job.

Maximum visuality is the goal of much modern technical communication in science, industry, and government. Nevertheless, as we have tried to show in this chapter, the rhetorical methodology and analytical skill valued in humanistic learning applies as much to visual as to verbal communication and can be used to produce unified texts aimed at getting a job done.

References

Amsden, D. (1980). Get in the habit of editing illustrations. *ITCC proceedings*.

Amsden, D. (1982). Exercise your visual thinking. *ITCC proceedings*.

Aristotle. (1982). *The "art" of rhetoric.* J. H. Freese (Trans.). Cambridge, MA: Harvard University Press.

Arnheim, R. (1969). *Visual thinking.* Berkeley, CA: University of California Press.

Aumuller, B. (1971). Industrial photography. In S. Jordan, J. Kleinman, and H. Shimberg (Eds.) *Handbook of technical writing practices*, Vol. 2 (pp. 803–818). New York: Wiley and Sons.

Barzun, J., & Graff, H. F. (1970). *The modern researcher.* New York: Harcourt, Brace & Jovanovich.

Beatts, P. M. (1959). Use your reader's eyes. *IRE Transactions on Engineering Writing and Speech.* EWS-2, 6-11.

Bertin, J. (1981). *Graphics and graphic information processing.* W. Berg and P. Scott (Trans.). New York: Walter de Gruyter.

Cox, B., & Roland, C. (1973). How rhetoric confuses scientific issues. *REE Transactions on professional communication.* PC-16, 140–142.

Englebret, D. (1972). Storyboarding—a better way of planning and writing proposals. *IEE Transactions on professional communication.* C-15, 115–118.

Enrick, N. L. (1972). *Effective graphic communication.* New York: Auerbach.

Freese, J. H. (Trans.) (1982). *Aristotle: The art of rhetoric.* Cambridge, MA: Harvard University Press.

Green, R. (1985). The graphic oriented (GO) proposal primer. *ITCC proceedings*.

Gross, A. (1983). A primer on tables and figures. *Journal of Technical Writing and Communication, 13,* 33–55.

Hairston, M. (1982). The winds of change: Thomas Kuhn and the revolution in the teaching of writing. *College Composition and Communication, 33,* 76–88.

Holtz, H. (1979). *Government contracts: Proposalmanship and winning strategies.* New York: Plenum.

Holtz, H. & Schmidt, T. (1981). *The winning proposal: How to write it.* New York: McGraw-Hill.

Huff, D. (1954). *How to lie with statistics.* New York: Norton.

Hughes, T. (1985). The graphic truth. *American Heritage of Invention and Technology,* **1,** 28–30.

Kinneavy, J. (1971). *A theory of discourse.* New York: Norton.

Lefferts, R. (1981). *Elements of graphics: How to prepare charts and graphs for effective reports.* New York: Harper and Row.

MacGregor, A. (1978). Selecting the appropriate chart. *IEEE Transactions on Professional Communication,* PC-21, 106–107.

McKim, R. H. (1980). Thinking by visual images. In W. S. Anderson & D. Cox (Eds.), *The technical reader* (pp. 104–114). New York: Holt, Rinehart and Winston.

Magnan, G. (1971). Industrial illustrating and layout. In S. Jordan, J. Kleinman, H. Shimberg (Eds.), *Handbook of technical writing practices,* Vol. 2 (pp. 735–800). New York: Wiley and Sons.

Mathes, J. C., & Stevenson, D. (1976). *Designing technical reports.* Indianapolis, IN: Bobbs-Merrill.

Mattingly, W., & Whitesell, M. (1985). Comprehending complex patterns of information using visual techniques. *ITCC proceedings.*

Maynard, J. (1982). A user driven approach to better user manuals. *IEEE transactions on professional communication,* PC-25, 16–19.

Murgio, M. (1969). *Communications graphics.* New York: Van Nostrand Reinhold.

Oslund, R. (1962). Brochuremanship versus cost. *Data,* **6,** 28–29.

Rockett, F. (1959). Planning illustrations first simplifies writing later. *IRE transactions on engineering and speech,* EWS-2, 56–61.

Rude, C. (1985). Format and typography in complex instructions. *ITCC Proceedings.*

Scofield, R. J. (1977). Can graphics do the entire communication job? Paper presented at the 24th International Technical Communication Conference, Chicago, May 11–14. In *24th ITCC Proceedings* (pp. 174–176). Washington, DC: Society for Technical Communication.

Skees, W. (1982). *Writing handbook for computer professionals.* Belmont, CA: Lifetime Learning Publications.

Stratton, C. (1984). *Technical writing: process and product.* New York: Holt, Rinehart and Winston.

Titen, J. (1980). Application of Rudolf Arnheim's visual thinking to the teaching of technical writing. *The Technical Writing Teacher,* **7,** 113–118.

Tracey, J. (1983). The theory and lessons of STOP discourse. *IEEE transactions on professional communication,* PC-26, 68–78.

Tufte, E. (1983). *The visual display of quantitative information.* Cheshire, CT: Graphics Press.

Turnbull, A. T., & Baird, R. (1975). *The graphics of communication* (3rd Ed.). New York: Holt, Rinehart and Winston.

Weisman, H. (1980). *Basic technical writing.* (4th Ed). Columbus, OH: Charles E. Merrill.

8

How Can Technical Writers Effectively Revise Functional Documents?

MARY DIELI

Revision is the implied subject of several discussions in this anthology. In her essay, Mary Dieli outlines specific revision techniques writers can apply to their functional documents to ensure that the intended audiences understand a document's purpose. Typically writers rely on revision aids such as style manuals, readability tests, or updated information about a document's subject. Or, if they have a finished document, writers can ask potential users to assess the document style and organization; they can then revise the document to address problems identified by these users. Most "revision sources" are time-consuming and expensive, and they cannot ensure that the intended audience will be able to apply the information in a text to a specific problem, the ultimate criterion for any functional document.

Although the most reliable way to predict a document's success is to test it with its intended audience, few technical writers have this luxury. Dieli suggests another way to examine documents from a user's point of view: designing and applying revision filters, document evaluation methods based on an analysis of the expected behavior of the document's users. A revision filter identifies the principle of user behavior for specific text and allows a writer to construct a set of steps to evaluate a document in terms of the principle. For example, functional documents are seldom read sequentially by a homogeneous audience. More often individual readers skip to specific sections that they can apply to their immediate problem. Thus, a principle for user behavior is a reader must be able to find information easily by using any part of a document as reference. From this assessment, Dieli suggests a test writers can use to evaluate a document with the reader's goal of access in mind.

150

The Need for Efficient, Audience-Focused Document Evaluation Methods

To produce clear documents, writers use various revision aids. For example, style guides help a writer revise a text so that it meets accepted language standards and writing conventions. When revising technical drafts, writers are often directed by information on the product they are writing about. As they get more information about it, they change the draft. In addition, there are often production constraints on writing, such as length, page size, and format. To add to these prescribed revisions, writers call on their own experience and knowledge about what seems to work and not work for readers.

Meeting language standards, writing conventions, and accurate product descriptions helps produce a technically correct draft. But none of these text-focused revision aids ensures that the audience's needs are met. The ultimate test of a functional document is whether people can use it successfully. To help meet this criterion for success, we need audience-focused revision aids that encourage us to look at the document from the document user's point of view, to assess and revise writing based on knowledge about user behavior.

One way to discover how well a document reads is to try it with members of its intended audience. We can then revise based on the document's ability to meet users' needs. But while user testing provides the most complete and accurate record of how well a document works, paradoxically it demands a finished document. Consequently, user testing can only take place late in the document design process. This reveals a gap in current document evaluation methods. We know how to apply text-focused revisions early in the document design process, but this is not enough. We also need the information that user tests can give us; still we can't wait until a document is finished. How then do we produce an accurate user perspective about a document as it is being written, rather than after it is written? A major goal of this chapter is to describe a method, revision filters, that answers this question. Revision filters allow writers to evaluate their documents faster and cheaper.

This chapter, in fact, has three goals. First, it discusses current methods of document testing, explaining how they are used to evaluate and revise documents. Second, it describes some specific problems users commonly have with functional documents. Finally, it discusses an evaluation method called *revision filters*. Revision filters close the evaluation gap between the early need for document evaluation and the need for user testing a finished document. Based on theories about how readers understand and use documents, these revision filters act like glasses that enable the document writers to see from the users' perspective. Looking at a draft in any stage of its development, a writer can apply these filters to predict certain kinds of user responses. The writer then can revise to eliminate document features that would cause problems for the audience.

Relevant Research: Current Evaluation Methods in Document Design

The general differences between text-focused and audience-focused tests, the two major types of document tests, can be summarized in cost–benefit analysis terms. Text-focused tests are cheaper to use than audience-focused tests because they are given to a large sample of document users. Audience-focused tests deal with a small sample of users because they are administered to one subject at a time. On the other hand, audience-focused tests, which allow the writer to get information directly from the audience, are one step closer to an audience model than text-focused tests; the text-focused tests are often based on user behavior, but the audience-focused tests measure user performance directly.

The distinction between the audience-focused and text-focused tests is not absolute. These tests represent the opposite ends of a continuum, with text focused representing one approach and audience focused the other. However, the different approaches are significant, and the categories help show the methods' strengths and limitations.

Text-Focused Measures: Readability Formulas

Readability formulas are the text-focused end of the methods' continuum. Common formulas measure two attributes of a text—a sentence's length and the prevalence of common words—and score the difficulty of the text based on those measurements. The prevalence of common words in the text is usually measured either by counting the number of syllables (because multisyllable words are usually less common) or by comparing the words to a dictionary of common words.

Although it is true that a text that receives a high readability score will likely be hard to understand, the reverse is not necessarily true. An incomprehensible text can get a low score on a readability test because these formulas measure individual words and their combinations. Readability formulas do not measure many other factors, such as content, organization, and style, that influence how well a text can be understood. The pros and cons of readability formulas have been researched extensively (e.g., Duffy & Kabance, 1982; Felker, 1980; Klare, 1983; Redish, 1979; Siegel, Federman, & Burkett, 1974; Stevens, 1980) as have the reader-based factors influencing readability (e.g., Frase, 1981; Holland, 1981; Kintsch & Vipond, 1978).

Text-Focused Measures: The Cloze Test

The cloze test was designed to measure readability by matching a text to its audience (Taylor, 1957). In the cloze test, blanks are substituted for words at regular intervals (every fifth word, for example) and the reader is asked to fill in the blanks (Bormuth, 1966; Panackal & Heft, 1978; Siegel, Federman & Burkett, 1974).

In addition to being used by professional writers, the cloze test is sometimes used by teachers to measure the readability of textbooks and, as Ken Davis suggests, by writing students in class to measure the readability of their drafts (Davis, 1982). According to Davis, students prepare the last 100 words of their drafts for the cloze test and then exchange the drafts with their peers. When they reach the blanks in the last 100 words, the students list their guesses on a separate sheet of paper. The students' cloze tests pinpoint text problems such as inappropriate shifts in person (e.g., "one" to "I"), unclear anaphoric references, poor transitions (e.g. "but" instead of "and"), and weak diction. More importantly, Davis makes the point that this application of the cloze test makes students more aware of their readers' needs.

However, Davis shows the limits of cloze analysis. First, the cloze test identifies local, that is sentence, problems rather than global, or text, problems. Although the students' guesses are based on the expectations they form while reading the entire text, their focus on weakness tends to be within a sentence or, at most, an intersentence connection. The test does not encourage students to consider problems like poor organization or poor thesis.

Intermediate Measures: Style Guides and Computer Programs

Style and design guidelines and computer programs for test analysis are classified as text focused; however, those that are based on reader behavior exemplify the distinction between text-focused and audience-focused tests. For example, several recent document design guidelines are grounded in basic and applied research about readers' reactions to text features (Felker et al., 1981; Felsenfeld & Siegel, 1981; Siegel & Glasgoff, 1981; Siegel, 1979). Other computer programs for English text analysis, notably those that make up the Writer's Workbench software, analyze document features ranging from spelling and punctuation to organization and style based on conventional writing standards and on psychological research in text comprehension (Frase, 1983; Gingrich, 1983; Kaufer, Steinberg, & Toney, 1983; Macdonald, 1983). These methods, generated by taking audience behavior into account, are classified as text focused because they give writers information about the text and not about the audience.

Intermediate Measures: Peer Review

Peer review involves asking readers to comment on a draft of the document (Cole & Cole, 1981; Cole, Rubin, & Cole, 1978; U. S. Department of Commerce, 1984). Generally peer reviewers are not members of the document audience and do not actually use the document as intended, so their criticisms have limited value. Their criticisms can vary depending on how the reviewers approach the task: they can limit their comments to matters of style and text conventions, for example, or they can try to simulate the audience and comment from the audience's perspective.

Audience-Focused Measures: Reader Protocols

Reader protocols are on the audience-focused side of the continuum. In a 1981 study, Swaney et al. used reader protocols to edit an insurance form after tests showed that expert editors were unable to improve the form (see also Flower, Hayes, & Swarts, 1980). Subjects giving reader protocols read and translate the text but do not use it as the intended audience will. However, since reader protocols let writers monitor the reading process, they clarify how some aspects of the document will work.

Audience-Focused Measures: Performance Tests and User Protocols

Performance tests are on the audience-focused end of the test methods continuum. In document design, performance testing means testing a document by observing a person from the intended audience using the document. During the test, various performance measures are taken, such as number and types of errors the user made, time necessary for the user to complete a task or section, and whether the user completed tasks correctly (Felsenfeld & Siegel, 1981; Goswami et al., 1981; Siegel, 1979).

Performance tests can indicate where problem areas in a document are, but the tests can't identify what in the document really caused the problem. User protocol tests are an enhanced version of performance tests that can supply that missing information because they add thinking-aloud protocols to the performance test (Atlas, 1981; Ericsson & Simon, 1984). During the controlled conditions of a user protocol test, users who represent a document's intended audience read the document and express their reactions aloud while using it. As they work, the users are tape-recorded and observed, preferably by the document's writer. The tapes and notes from the protocols are carefully analyzed, and the document is revised based on what the analyses show about the users' needs. The revised document is then tested on two or more new users. This process is repeated until users encounter no major problems.

The advantages of user protocols are that (1) few problems escape the attention of the writer; (2) the reasons for the problems are pinpointed, and therefore the test can lead to a concrete revision plan; and (3) the writer learns many things about users that can be applied to other documents. The disadvantages are that (1) they are time-consuming and, therefore, expensive; (2) writers need training and practice in order to construct the tasks for users, conduct the protocol properly, and make the best use of the wealth of information obtained; and (3) the setting for the tests is somewhat artificial for the user.

Recommendations

Translating Audience Information into Revision Heuristics

As the above survey of evaluation methods points out, the surest way to find out if a document works for its audience is to test it with members of that audience. This is especially true for documents with instructions or procedures to follow in addition to explanatory text. Document testing with users not only points out what revisions need to be made, they also show the special set of task demands users face which may affect the way they use the text. For example, when I tested Apple Computer's Lisa Reference Guides, my goal was to help users complete a task by moving through the Reference Guides independently. To be successful, technical writers need to design functional documents to respond to the special problems using such documents create.

Revision Filters

Revision filters are important as a document evaluation method because they are based on user behavior. Developed from user test data, each filter describes reader response to one document feature (for example, symbolic information) and presents a test writers can use to predict user problems.

User testing the Lisa documentation, I saw four major areas where users commonly had problems:

- Interpreting symbolic information.
- Knowing when to switch among the ways one responds to text.
- Finding needed information.
- Choosing the correct solution for a problem.

Below, I explain each problem in more detail, quoting from user protocols to illustrate each problem. Then I present the filters I wrote to alert writers to each user problem.

Problem 1: Interpreting Symbolic Information

Functional documents typically contain symbolic information that readers must actively interpret as they use the document. Audiences have three types of problems with symbols: (1) they don't notice them, (2) they misinterpret them, and (3) they perceive a symbol as significant when it is not intended to convey any instructional meaning. For example the following protocol quotation shows a user trying to revise a document in *LisaWrite* (a word-processing program) so that the page number appears along the right side of the screen. He is reading down (scrolling) through the ruler looking for its footer symbol. Because he doesn't know what the symbols mean, he makes the wrong assumption and can't complete the task. (The written text is shown in italics.)

> OK now I'll scroll down . . . now I wanted something exotic happening . . . I don't know how long it goes oops oh no it was 8 it turns out not a funny little mark . . . an 8 inch page seems a little unreasonable to me . . . What I'm doing here is I'm sitting waiting for some mark to come oops now wait a minute there is a little mark at the 10 . . . so maybe have a ten inch page Does that mean that? Oh yeah I see here it says 10 . . . ah oh and you can see the footer marker. *You select a point of insertion beside the footer marker.* Now let's see if that is the footer marker let's continue scrolling til we see . . . I see this little mark B and then there's a mark F and um I don't know what those letters mean. *Choose the page insert number from the page layout. . . . Select the point of insertion beside the footer marker. A blinking bar goes where inserted text will go.* Inserted text? Oh OK inserted text is indeed the uh . . . is indeed I suppose well I'm going to assume that B for bottom rather than F for footer is what we're talking about here oops sorry there little marker I'm going to put you right roughly in the middle here I don't know if I'm supposed to measure I'm going to assume not all right let's hope our blinking marker will appear . . . it did not.

Problem 2: Knowing When to Switch Responses

Functional documents typically present information to users in several different forms: (1) procedures (i.e., step-by-step instructions), (2) explanations (e.g., definitions, examples, and other types of text to read but not to act on), and (3) symbols (e.g., illustrations, punctuation, type styles, bullets, arithmetic symbols, directions about how to proceed in the document). While using a functional document, the reader must continually switch the way he is responding to the information a writer gives him, depending on the writer's purpose.

Below is a protocol quotation of a user looking for instructions about how to copy a formula from one cell to a range of cells. Notice her confusion as she tries to determine whether she is reading an explanation or a procedure.

> This is still an awful lot of text even before telling me what I need to do we should get to what I have to do. I can't you know I can't remember . . . it's just a lot of explanation without much doing. All right. *Suppose the formula in B2 is A2 plus C2.* OK here this seems more procedural now. *And you copy the cell using paste adjusting the range B4 B6. Suppose further that you want the first coordinate in each formula to be relative and the second to stay the same. The sample cell in the dialog box is B4. LisaCalc interpolates the formula as saying add the cell to the left of B4 and the cell to the right of B4. This translates into the formula A4 plus C4 which is shown in the dialog box.* This format here on this page is deceptive too because it it looks like it's . . . it's set up to imply that it's going to . . . um . . . be procedural with step-like you know 1 2 listing a formula but it's still very descriptive it's not it's not operational. I'm sick of just listening to their lecture here. *Since you have decided you want the second coordinate to stay the same, you click the box.* OK here now it's telling me to do something . . . the whole language has changed here. *You click the box above C4 and it changes to C2. After you click OK a similar formula is copied to every cell in the range A4 A6.* OK that's what I want to do. Now what I need to find out is how you actually let it know that um what the range is . . . OK . . . so I have to find that out . . .

Interestingly, the text that the user read is an example in an explanation section, not a procedure. But the example was presented as a series of numbered steps, and, as the user noted, some of the language was procedural rather than descriptive, signaling her to act. The writer's style prompted the user to assume that some of the text she was reading was meant to tell her how to do something when it was only intended to describe. This type of mistake can cause readers difficulty because examples usually don't contain enough procedural information to enable readers to perform a task correctly. For users to switch among the ways they respond to text, writers need to evaluate their documents to ensure that each piece of information clearly signals its purpose, both verbally and visually.

Problem 3: Finding Information
Functional documents are often not meant to be read sequentially from front to back; instead, the audience is expected to move around within a document depending on what information they require. In such docu-

ments, access to the appropriate information becomes a reader's primary goal.

The following question shows a user looking for a procedure in the table of contents. What she needs is the "Copy with Paste Adjusting" procedure located under the subheading "Edit." During her search, she has two problems: (1) she does not understand what some of the entries mean and (2) she makes an incorrect assumption about which subheading ("Calculate," "Edit," or "Format") the procedure is listed under.

> OK so now I've got to see if I can copy a formula in one cell . . . C14 B14 cell through M14 so that I can figure out so that I can figure out and display the number of installers needed . . . figure out and display . . . OK . . . Alright the table of contents, what does it tell me. *LisaCalc concepts, pointers, selecting, scrolling, about menus, dialog boxes, alert boxes, LisaCalc work flow, starting a new document* . . . *What's in chapter 2, Calculate, circle missing values, dates, formulas* . . . *formulas, circle missing values.* I don't know what that means. *functions* OK I need to copy the formula out of B14 into C14 through M14 so that they would operate on that information *recalculation, Edit* I'm not doing *Edit* I'm not doing *Format* I'm doing *Calculate* OK.

In fact, the procedure she wanted was under the subheading "Edit." After 37 minutes spent searching the table of contents and reading some inappropriate sections, she gave up on the task. She never found the correct procedure.

Problem 4: Matching a Solution to a Problem

Procedures are the basic information type in some functional documents. To complete a task, the document users must find and follow a procedure. However, before they can use the procedure, they must decide whether the feature they have looked up is the appropriate one.

The user in the following protocol excerpt has trouble understanding the entries in the table of contents but, after some searching, finds the correct entry, "Copy with Paste Adjusting." However, in the middle of trying the procedure, she sees two other related options, "Paste" and "Paste Values" displayed next to "Paste Adjusting" on the screen and becomes unsure about whether "Paste Adjusting" is the option she wants to use. She then reads most of the explanatory text trying to decide among the three.

> Oh this looks good. OK *to copy* select the cell that I want to copy. I think it's selected but I'll select it again . . . oops that did something that made me think then think I was going to insert something so I'm going to click away. Now I'll select it . . . OK select the cell I want to copy. *Choose copy from the edit menu* . . . OK . . . *Now select the new location.* It says singular I'm just going to try to copy it once for starters . . . so I select it. Now from the

edit menu I choose paste adjusting and I get a dialog box . . . um I think this is what I want maybe I want really paste . . . 13 . . . I'm going to hold this down while I read, I went too fast . . . *Copy cells with paste. . . . If you choose paste, the entire cell including its format, protection, value and formula is copied. The formulas are automatically updated so that they refer to the new cells not the old cells. References to formulas outside the range are not changed. Formulas outside the range being copied are not affected. See . . . example . . . If you choose paste values* it's not OK I don't want that because I want the formula. I think I want paste. Let's see what happens . . . that doesn't look reasonable because I got the same value . . . yeah it did . . . sigh . . . um . . . hmm . . .

Experimenter: What are you thinking?

I'm thinking that I'm used to the . . . um . . . I'm going to go I'm thinking that I misunderstood what paste meant . . . *If you choose paste, the entire cell including its format, protection, value and formula is copied.* OK so that's maybe the formula really did . . . let me select this and see if I get a formula out . . . oh I got a formula all right but it has b's in it instead of c's . . . OK. *When you copy a range of cells some of the cells in the range might have formulas that refer to other cells in the range.* um OK so I didn't use a range so maybe that um maybe that mattered. Let me see about paste adjusting it looked like I could do stuff . . . um . . . to the formula. Let me read this just to be thorough. *If you choose paste values* the formula is not copied that's clearly not what I want . . . um stuff about the clipboard I don't care about that um . . . oh *A single cell can be copied to a range of cells. A range can have any size or shape* . . .

What this user apparently needed was a brief introduction to those three procedures that explained what each one was, when to use each, and what the results of using each will be. She could then have matched her task goals with the stated results of each of the three choices and chosen accordingly.

Translating Audience Information into Document Evaluation Heuristics: Revision Filters

Identifying common user problems is only the beginning. The next step is to translate these problems into a form that gives writers a picture of user behavior before a document is ready for user testing, and so allows writers to revise with the user in mind earlier in the document design process. I developed a filter to respond to each of the problems outlined above. Each filter has two parts: the first is a principle of user behavior, described and with examples, and the second is a set of steps for writers to follow to evaluate their documents, or representative sections of them, in terms of the principle. (All four filters are in the appendix at the end of this chapter.) As an example, one filter, about symbolic information, is shown in Figure 8.1.

Principle

Functional documents typically contain much symbolic information. Any of the following may be interpreted as symbolic information:

- Headings.
- Page headers and footers.
- Sequence markers (e.g., bullets, numbers, dashes).
- Changes in color.
- Changes in type style (e.g., underlining, bold, italics).
- Changes in type size.
- Rules, boxes, arrows.
- Arrangement of items on the page.
- Pictures, illustrations.
- Special characters.

Much of the symbolic information in functional documents contains unstated instructions for readers. For example, changes in type size may cue readers about the relative importance of information, or "^S" may mean "Hold down the 'ctrl' key and hit the 's' key." To use a document successfully, readers must notice and correctly interpret its symbolic content.

The following test asks you to notice all the symbolic information in a section of your document and to evaluate it in terms of readers' needs.

Test to Be Used by the Writer

1. Using the list above as a checklist, read through a representative section of your document and circle all the symbolic information.
2. Use the guidelines below to evaluate each item circled:

   - What is the symbol's unstated instruction to the reader?
   - Does the symbol need to be identified and explained?
   - Is the symbol unnecessary and so potentially confusing?
   - Are there any alternatives that would work better?

3. Finally, review the symbolic information in the whole document and revise so that the symbolic information is consistent throughout the document.

Figure 8.1 Symbolic Information Filter

The filters have some features that make them valuable revision aids for writers of functional documents. First, they are based on theories of how readers use documents rather than on standards that may or may not be appropriate to a given document. Second, they begin to fill the evaluation gap with efficient and effective revision aids that can focus the writer on user behavior early in the document design process. And finally, they account for some special features of functional documents that redefine document readers to be document users. Writers can apply the filters to their documents to revise them with the document users in mind.

Testing the Revision Filters

To learn whether the filters, with their audience information, can help writers evaluate documents, I asked professional writers to test them. The purpose of the test was to compare the accuracy of each method as a predictor of problem areas in computer documentation.

> Signal detection analysis is one way to measure the accuracy of different document evaluation methods. Signal detection is often used in hearing experiments where subjects have to distinguish the presence or absence of a particular auditory signal (or tone) from a background of other noise. The subjects are asked to detect whether the tone is present or not in a large number of instances. A subject with sensitive hearing would be a good detector; a deaf subject would obviously be a poor detector. (Carey, 1985, p. 8)

For the research reported here, the test methods are considered to be predictors of audience problem areas in the documentation; audience problem areas are the signals that the methods are attempting to detect.

To test the filters, 30 writers were asked to read sections of two computer manuals and to evaluate them, describing problems they thought the audience might have with the documents. The writers were randomly assigned to three groups of 10. One group of writers used a set of four "traditional" writing guidelines, one group used the revision filters, and one group worked with no prompts. All the writers gave verbal protocols as they worked. Ten members of the documents' intended audience also read the manual sections, translating the content and thinking aloud as they did. With the verbal protocol data, I compared the document problem areas that the readers pinpointed with the writers' predictions.

Figure 8.2 illustrates the accuracy of the writers' predictions. (The y axis shows the probability of a hit, and the x axis shows the probability of a false alarm. The diagonal line shows where the probability of a hit equals the probability of a false alarm.) In signal detection analysis terms, the unprompted writers' probability of a hit was .718, and their probability of a false alarm was .515. The writers working with filters had a .849 probability of a hit and .52 probability of a false alarm, and the writers working with guidelines had .346 and .283 probabilities of hits and false alarms, respectively.

To look at the results in terms of accuracy, the writers working with filters were the best predictors of problem areas in the manuals, and the writers working with guidelines performed the worse. Signal detection analysis shows the writers working with filters to be more thorough detectors of problem areas than the writers working without prompts; writers with filters have a greater percentage of hits (84.9%) than the unprompted

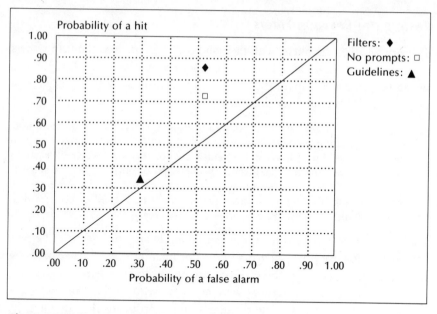

Figure 8.2 Writers as Detectors of Problem Areas in Two Computer Manuals

writers (71.9%). The writers with guidelines are the least thorough of the three groups with a 34.6% chance of hits.

The points' distance from the diagonal on the signal detection graph shows that the writers working with filters are more sensitive detectors of problem areas than both the unprompted writers and the writers working with guidelines. Given that there is such a difference, the next question to ask is how large that difference is. A test confirmed that both the writers working with filters and the writers working without prompts performed significantly better than chance whereas the writers working with guidelines did not. (For both groups, significance is .025, p > .05.) However, an analysis of variance did not show a significant difference across groups. Thus, the signal detection analysis shows promise that there is a difference in performance across groups, promise that filters improve both writers' thoroughness and sensitivity.

With such positive initial results, the next step in this research is to compare the writers' predictions with problems that users have while using the documents to help them solve problems on the computer. However, the message seems clear: the use of audience heuristics can help writers produce readable functional documents.

APPENDIX   Revision Filters

Filter 2: Access

Principle

Functional documents are often not meant to be read sequentially from front to back; readers are expected to move around within a document depending on what they are looking for and what they are working on. Many functional documents demand that readers look up information to complete an action or answer a question. With such documents, access of the correct information becomes the reader's primary goal.

One reason readers may have trouble finding the information they need is that tables of contents and indexes may contain jargon terms unfamiliar to the average reader. However, a blanket jargon to nonjargon conversion will cause access problems for users familiar with, and looking up, jargon. To begin to solve the access problem, then, we need to revise with the broad audience in mind.

Readers may also have trouble finding what they need because the table of contents and index do not contain enough information. The table of contents may list only major headings, not subheadings, and the index may simply not contain enough entries.

The following text asks you to evaluate your document's table of contents and index with the reader's goal of access in mind.

Test to Be Used by Writer

Type of Information

- Read through the table of contents and circle all jargon terms, specialized vocabulary that may not be familiar to the average reader.
- For each jargon term, think of a nonjargon term that describes the same activity or feature.
- Consider revising the table of contents so that a jargon term is followed by its nonjargon synonym (separated with colons or parentheses) and revising the index to contain both jargon and nonjargon terms.

Amount of Information

Compare the levels of headings in the document with the levels listed in the table of contents. If there is a difference, consider expanding the table of contents by including subheadings.

Filter 3: Switching Response

Principle

Functional documents typically present information to the reader in a number of different forms. For example:

- Procedures (step-by-step instructions).
- Explanations (e.g., definitions, examples, and other types of text to read but not to act on).

• Symbols (e.g., illustrations, punctuation, type styles, bullets, arithmetic symbols).

Information in each form asks the reader to respond in a certain way. We can think of each form as ideally containing embedded directions that signal the reader to respond appropriately: to act, to read, to learn, or just to notice. For example, when reading a procedure the reader should know "I am to do [an action]." Explanations should signal "learn [this information]."

Since readers must continually switch the way they are responding to information depending on its purpose, we must design the information to signal its purpose clearly. If the differences among information forms are not clear, readers may respond inappropriately, trying to use an explanation as a procedure, for example, or ignoring or misinterpreting symbolic information.

The following test asks you to focus on the different forms of information in your document, to evaluate their intended purpose, and to evaluate how well your document signals the reader to respond appropriately.

Tests to Be Used by the Writer

• Read through a representative section of your document and categorize information according to its form (as procedures, explanations, or symbols).
• Look at the boundaries between forms and find the visual cues (changes in, or new, graphic design elements) or verbal cues (headings, directions) that signal the reader to switch processing.
• Consider whether or not such "switch points" signal the reader clearly and adequately.

Filter 4: Matching a Solution to a Problem

Principle

Procedures, step-by-step instructions for doing something, are the basic information type in some functional documents. In order to complete a task, the manual user must find and follow a procedure. However, first they must decide whether the procedure they have looked up is the appropriate one. To help the user decide, procedures need to be reinforced with definitions and explanations.

In procedural documents, we can think of the above three information types as doing the following:

• Definition: defines what the procedure accomplishes and when (under what conditions) it should be used.
• Explanation: describes the procedure in context, with examples.
• Procedure: step-by-step instructions.

If procedures are preceded by definitions and explanations, then readers can look up a procedure, read a definition to ascertain that the procedure is what they want, read an explanation to further confirm that the solution chosen matches the problem and to better understand procedure details, and then use the procedure.

Test to Be Used by the Writer

• Find a procedure.
• In the related text, find a definition. This should contain a brief description of what the procedure accomplishes and when it should be used.

- Find related explanatory text. This should contain examples and context for use.
- Write a definition and an explanation if they are not there.
- If necessary, reorganize information so that definitions and then explanations precede procedures.

References

Atlas, M. A. (1981). The user edit: Making manuals easier to use. *IEEE Transactions on Professional Communication, 24*(1), 28–29.

Bormuth, J. (1966). Readability: A new approach. *Reading Research Quarterly, 1,* 79–132.

Carey, L. (1985). Statistics for protocol analysis. Unpublished paper.

Cole, J. R., Cole, S., with the Committee on Science and Public Policy, National Academy of Sciences. (1981). *Peer review in the National Science Foundation: Phase two of a study.* Washington, DC: National Academy of Sciences Press.

Cole, S., Rubin, L., & Cole, J. R. (1978). *Peer review in the National Science Foundation: Phase one of a study.* Washington, DC: National Academy of Sciences Press.

Davis, K. (1982). The Cloze test as a diagnostic tool for revision. In R. Sudol (Ed.), *Revising: New essays for teachers of writing* (pp. 121–126). Urbana, IL: ERIC & NCTE.

Duffy, T., & Kabance, P. (1982). Testing a readable writing approach to text revision. *Journal of Educational Psychology, 74*(5), 733–748.

Ericsson, K. A., & Simon, H. A. (1984). *Protocol analysis: Verbal reports as data.* Cambridge, MA: MIT Press.

Felker, D. B. (Ed.). (1980). *Document design: A review of the relevant research.* Washington, DC: American Institutes for Research.

Felker, D. B., Pickering, F., Charrow, V. R., Holland, V. M., & Reddish J. C. (1981). *Guidelines for document designers.* Washington, DC: American Institutes for Research.

Felsenfeld, C., & Siegel, A. (1981). *Writing contracts in plain English.* St. Paul, MN: West.

Flower, L. S., Hayes, J. R., & Swarts, H. (1980). *Revising functional documents: The scenario principle.* Technical Report No. 10. Pittsburgh, PA: Carnegie-Mellon University.

Frase, L. J. (1981). Ethics of imperfect measures. *IEEE Transactions on Professional Communication, 24*(1), 48–50.

Frase, L. J. (1983). The UNIX Writer's Workbench software: Philosophy. *The Bell System Technical Journal, 62,* 1883–1890.

Gingrich, P. M. (1983). The UNIX Writer's Workbench software: Results of a field study. *The Bell System Technical Journal, 62,* 1909–1921.

Goswami, D., Redish, J. C., Felker, D. B., & Siegel, A. (1981). *Writing in the professions.* Washington, DC: American Institutes for Research.

Holland, V. M. (1981). *Psycholinguistic alternatives on readability formulas.* Document Design Center Technical Report No. 12. Washington, DC: American Institutes for Research.

Kaufer, D. S., Steinberg, E. R., & Toney, S. D. (1983). Revising medical consent forms: An empirical model and test. *Law, Medicine & Health Care,* **11**(4), 155–163.

Kintsch, W., & Vipond, D. (1977). Reading comprehension in educational practice and psychological theroy. In L. G. Nilsson (Ed.), *Perspectives on memory research* (pp. 329–365). Hillsdale, NJ: Erlbaum.

Klare, G. R. (1983). Readability indices: Do they inform or misinform? *Information Design Journal,* **2**(3&4), 251–255.

Macdonald, N. H. (1983). The UNIX Writer's Workbench software: Rationale and design. *The Bell System Technical Journal,* **62**, 1891–1908.

Panackal, A. A., & Heft, C. S. (1978). Cloze techniques and multiple-choice technique: Reliability and validity. *Educational and Psychological Measurement,* **38**, 917–932.

Redish, J. C. (1979). Readability. In D. A. MacDonald (Ed.), *Drafting documents in plain language,* Handbook A4-3034 (pp. 157–174). New York: Practising Law Institute.

Siegel, A. I. (1979). Drafting simplified legal documents: Basic principles and their application. In D. A. MacDonald (Ed.), *Drafting documents in plain language* (Handbook A4-3034, pp. 175–252). New York: Practising Law Institute.

Siegel, A. I., Federman, P. J., & Burkett, J. R. (1974). *Increasing and evaluating the readability of Air Force written materials.* NTIS AD-786-820. Lowry Air Force Base, CO: Air Force Human Resources Laboratory, Technical Training Division.

Siegel, A. I., & Glasgoff, D. G., Jr. (1981). Case history: Simplifying an apartment lease. *Practicing Law Institute,* 169–205.

Stevens, K. C. (1980). Readability formulas and McCall-Crabbs standard test lessons in reading. *The Reading Teacher,* **33**, 413–415.

Swaney, J. H., Janik, C. J., Bond, S. J., & Hayes, J. R. (1981). *Editing for comprehension: Improving the process through reading protocols.* Technical Report No. 14. Pittsburgh, PA: Carnegie-Mellon University.

Taylor, W. L. (1957). "Cloze" readability scores as indices of individual differences in comprehension and aptitude. *Journal of Applied Psychology,* **41**, 19–26.

U.S. Department of Commerce. (1984). *How plain English works for business.* Washington, DC: Government Printing Office.

9

How Can Technical Writers Be Effective Research Collaborators?

ELAINE ELDRIDGE

Technical writing, a relatively new profession, demands collaboration between the writers and the research staffs they support. Yet the very newness of the discipline encourages some researchers to discredit technical writers' contributions and to exclude them as integral members of the research team. For many scientists and managers, technical writers are sophisticated secretaries rather than skilled professionals. Researchers frequently call for a technical writer's expertise too late in a research project for the writer to be of assistance. Rather than understanding that a technical writer's expertise is communicating complex ideas to various audiences, skeptical technicians and scientists consider technical writers as editors who correct grammar—not as colleagues who shape the document's form and style.

As a writing consultant for private industry and a professor of technical writing, Elaine Eldridge outlines the elements required for successful collaboration between researchers and technical writers and recommends strategies technical writers should use to achieve this collaboration. The most significant factor for professional success is interpersonal skills: technical writers must not only be able to write persuasively to a variety of audiences, they must also be able to understand the disciplines of the various professionals with whom they work. This interpersonal skill leads to greater respect from the writers' colleagues and, subsequently, earlier involvement in the research project. When the technical writer becomes involved in a research project at its inception, the project's final report will synthesize scientists' research with the technical writers' skills.

Recognition and Technical Writers

Technical writers rarely work alone. After completing the degree, landing the job, or signing the contract, they must be prepared to work with others. Such collaboration almost always calls for a definition—sometimes a defense—of the technical writers' skills, purpose, and reason for employment because, unlike many occupations, technical writing is still establishing itself (see Chapter 12). Confusion over a technical writer's role as a research collaborator stems from two basic sources: first, the varied educational and employment backgrounds of most technical writers and, second, colleagues' misunderstanding or ignorance of the technical writer's contribution to research communication.

Educational and Employment Backgrounds

Although there have been "English for Engineers" courses for decades, as a university subject technical writing is comparatively new. It is only since the 1970s that technical writing has flourished in university curricula nationwide, and only more recently that undergraduate degrees in technical writing have been established (Connors, 1982; Gould & Losano, 1984). Most working technical writers come from related disciplines, such as journalism or English; some, from engineering and the sciences; and some technical writers' educations bear little relation to their work. Thus, not only have most technical writers had to learn on the job, they have had to define themselves without the expected aid of a recognized college degree. Although the new degree programs are a welcome effort to provide university training and recognition, they are too few and too recent to have established technical writing as a widely recognized occupation. Further, the comparative newness of these degree programs means that no generally accepted standards or requirements yet exist for a bachelor's degree in technical communication (Connors, 1982).

Technical Writers' Contribution

This lack of standardization in education necessarily leads to a lack of professional identity. In the eyes of managers and colleagues, "technical writers" can refer to anyone from secretaries who "edit" a manuscript (i.e., correct spelling and put in headings) to tenured research scientists who impeccably write up their own research. Without a widely known or accepted professional identity, technical writers face the second major source of confusion over their collaborative role: the scientists, engineers, and managers with whom they work—or try to work. Repeatedly, technical writers report frustration at not receiving full professional recognition, at not being an integral part of the research team, at not being included in management decisions (Zook, 1983). These problems are usually caused by colleagues not understanding what the writers can do for a research project or seeing them only as members of the technical support staff.

At worst, such misunderstandings result in the technical writers' being considered as little more than elaborate secretaries. Recently, for example, I spoke with the president of a large midwestern agricultural company about writing a series of technically oriented marketing brochures. I knew I was facing serious difficulties in professional collaboration when he thought it more important for me to meet the secretary in charge of typing and manuscript production than the scientists whose work the brochures would describe. More common is the technical writer who is a valued, but secondary, member of the research effort. For this writer, collaboration usually means being called in at the end of a project to revise or edit the final report already written by the research team.

Ideally, of course, technical writers are primary members of the research team: the writers' subject expertise, skill in synthesizing complicated ideas and communicating those ideas to a variety of readers, and ability to produce a document that will explain and validate the team's work make them essential to the research effort. In such cases technical writers are true collaborators, colleagues who are present from the opening brainstorming sessions through the development and completion of a project.

In other words, what "collaboration" means for working technical writers is inextricably linked to the chief question in professional collaboration: what is the technical writers' status with the scientists, engineers, managers, and administrators with whom they work? How these people view technical writers and how the writers respond to their views determine the degree of integration into the research team. Much of this book suggests solutions to problems in technical writing; this chapter suggests strategies for how you can cope with the challenges of *being* a technical writer, how you can define technical writing to sometimes skeptical or reluctant colleagues, and how you can establish yourself as a respected collaborator on the research and development team.

After a brief description of the status of current research on this question, the final section of this chapter considers what successful professional collaboration includes and identifies the elements required for such collaboration. The chapter concludes with a discussion of the problems writers face in a less than perfect working world and offers recommendations for solving them.

Relevant Research

The almost total lack of research specific to the question of the technical writer as collaborator underscores the first paradox writers face: they have been working for years at a profession still defining itself. The problem is not that the questions of professional identity confronting the technical writer are insignificant but that no one has thought of asking the questions. For example, Harris (1978) addresses the problem of expanding the definition of technical writing, but his underlying assumption is that writers don't need to expand their role and definition. Connors (1982) refers several times to a "growing professionalism" in technical writing, but his article focuses on the growth of technical writing instruction and increased self-awareness among technical writing teachers. Two other articles, by Power (1981) and Zook (1983), directly confront the responsibilities technical writers and editors face on the job. Power (1981, p. 139) describes a technical editor working closely with an engineering staff to help "start a writing project and then provide ongoing expert guidance until the report goes in the mail"; she also sees the editor as someone who can "design and present in-house writing courses," "coach individuals in effective writing techniques," and "supervise and manage a word-processing area." Although she stresses teamwork, Power awards priority to the engineers, whom she designates as "authors"; technical editors remain assistants. "To be a technical editor takes some diplomatic skills, as well as writing ability. . . . If everyone sticks to the job he does best, authors and editors can cooperate to get the work done promptly and correctly" (Power, 1981, p. 140). There is an inherent contradiction here. If these author-engineers stick to the job they do best, they will concentrate on engineering and let the editor-writers concentrate on the writing. Finally, in her interesting article describing the frustrations and satisfactions technical editors face as collaborators, Zook (1983) discusses working with management and researchers as well as the tremendous range of job duties writers face. But Zook, like Power, concentrates on the editor as an adjunct to the primary writer-researcher.

Recommendations

Interpersonal Skills

Although all of the writers mentioned above treat them with sympathy and respect, they do not see technical writers as professional collaborators or as primary members of the research team. But successful collaboration, while not easy, is entirely possible; it depends on three closely related factors:

- A technical writer's interpersonal skills.
- Colleagues' attitudes.
- Writer's length of involvement on the research project.

The first factor, interpersonal skills, is essential to any writer's success. Writers work with a wide range of people, not only the scientists on the research team but management, administration, accounting, other technical writers, the support staff, and departments such as graphics and public relations. Moreover, technical writers need to be sensitive to the needs of the various offices and departments with which the research team must communicate. If all we had to worry about was writing for other knowledgeable professionals, technical writing would be easy. But as the team's communication experts, technical writers are responsible for informing people who have a wide range of technical expertise and interest in the research: clients, customers, editors, state and federal officials, and review boards. Further, technical writers are almost always considered the team's experts on bibliography and documentation, which means that they are not only comfortable with computerized literature searching and the more traditional secondary research techniques but can work effectively with librarians and interview outside resource people as needed.

Related responsibilities usually include overseeing the production of technical documents. For in-house reports, document production increasingly requires familiarity with word-processing systems, including text editing and formatting systems; it can also require the writers to work closely with the support staff. Depending on the nature of the research and the size of the organization, a technical writer may be in charge of photography and graphics (or at least familiar with the capabilities and personnel of the graphics department).

The point is that these are not skills practiced in isolation—they almost always require working with other people. In one of the few studies that mentions this point, over half the respondents rated interpersonal skill ("defined as the ability to work well not only with writers but with all the other people involved in communication projects") as the most significant factor in professional success (Zook, 1983, p. 23). Yet as technical writers we are so often concerned with clarity and accuracy that we can forget that an abundance of writing expertise cannot make up for disgruntled or irritated colleagues.

That interpersonal skills and collaboration go hand in hand won't surprise any technical writer who has ever struggled to be accurate, clear, tactful, and meet a deadline—all at the same time. The interesting point is that these skills are more important for the technical writer than for the writer's colleagues. This is partly true because technical writers, as the newest members on research and development teams, frequently work harder than established members to prove their worth and acceptability. More significantly, the technical writer's work on the research team mediates the old split between scientific technology and humanistic communication. Predictably, then, the writer's collaborative responsibility is greater than the other team members' because the writer's contribution results from a synthesis—a *private* collaboration—of technical knowledge and communication skills. Even the name "technical writer" points to the fact that before we are fully integrated into the research team, we first perform an internal integration of our scientific and humanistic knowledge.

Collaborators' Attitudes

The second two factors required for successful collaboration, positive attitudes from colleagues and the writer's length of involvement on the research project, can be most clearly illustrated with an example. At the Fred Hutchinson Cancer Research Center in Seattle, I wrote an article with a biostatistician who specialized in cancer research (Polissar & Eldridge, 1982). Our subject was the potentially carcinogenic effects of asbestos in drinking water. I was familiar with the problem from having previously worked with him on an article for a professional journal and from having edited several E.P.A. reports on the same subject. Audience was of particular concern for us. Because the article's readers were a mixture of health science professionals, including practicing M.D.'s, research physicians, social workers, hospital administrators, epidemiologists, and public health specialists, the level of technicality had to be between other statistical experts and the well-educated "general" reader.

The final document was a synthesis of our writing and research skills. Each of us was responsible for reviewing current studies of the relationship between cancer incidence and asbestos in drinking water. In addition to the shared responsibility of subject knowledge, we had different responsibilities based on our specialized skills. Because he was the primary investigator, much of my colleague's field research was completed or nearing completion. My job, of course, required drafting and completing an article that would meet our requirements, satisfy our editor, and educate our target audience. To accomplish this I worked closely with my collaborator at all stages of the article's development. We decided what we wanted to include and carefully considered the needs, interests, and knowledge of our readers. To supplement the statistical information, I conducted a telephone interview with a city water department official, an expert on the asbestos–cement pipe that carried local water supplies; I also studied recent issues of the journal that had solicited the article and spoke briefly with its editor. Not surprisingly, my partner and I went through several drafts before we were finally satisfied with our text.

What made this project successful? Subject familiarity, research, and interviewing skills all played a part, but the key answers to this question are the second two elements of successful collaboration: *attitude* and *length of involvement*. First, each of us approached the job with confidence in our own abilities and a respectful appreciation of the other's expertise. I recognized my collaborator's knowledge of the carcinogenic effects of asbestos in drinking water was considerably greater than my own. More importantly, he recognized my technical communication skills. Our quite different abilities meshed to produce an informative article that pleased each of us. Second, I was involved with the article from beginning to end, from the time the editor first suggested it until publication.

"From beginning to end" is a crucial factor for successful collaboration as well as being a significant indicator of your colleagues' or employers' attitudes toward you. Briefly, "from beginning to end" means that the technical writer actively participates in a research project from its inception to an articulated plan to its testing to, finally, an article or report for other researchers, administration executives, funding organizations, the appropriate federal agencies, and whoever else has an interest in the research results.

A second brief example may clarify and emphasize the importance of length of involvement. The National Institute of Safety and Health (NIOSH) had announced funding availability for studies of cancer incidence among children whose parents had been exposed to known carcinogens. In collaboration with a biostatistician and an epidemiologist, I began reading NIOSH's research solicitation and proposal specifications, which suggested what research was feasible for their needs and our abilities. From there the team began, in bits and pieces, in meetings and telephone conversations, to draft the proposal. This included explaining and justifying the proposed research and showing how it fit the agency's specifications. I also worked with the researchers to develop a detailed and unbiased questionnaire for gathering the quantifiable information they would need. In this instance, I not only worked with the team until the proposal was ready to send but actually began working even *before* we had a definite idea in mind.

These examples describe an ideal collaboration on typical research projects. Fully collaborating technical writers participate in brainstorming and preliminary planning sessions and are present when the research team delegates responsibilities to its members. At this stage the technical writer must be familiar with current research and early drafts of the project. Although not an expert on the proposed research, the writer understands the project's purpose and methodology, can identify obvious problems, and makes pertinent suggestions. The writer stays informed as the research plan is refined and developed, continues to keep up with newly published research, writes progress reports and memos as needed, and keeps careful written records so that the final report is accurate. If thoroughly involved with the research from its beginning, the technical writer can produce an authoritative and comprehensive report on the research findings when the project is completed.

Before turning to some of the challenges the technical writers will meet in achieving success as a collaborator, I would like to comment on the obvious omission in this discussion: writing ability. No technical writer will survive without it, but writing ability is not strictly a *collaborative* skill. Any professional writers, not just technical writers, must be able to write. After their interpersonal skills gleam from polished use, technical writers face their second major paradox: reports are written alone.

Trouble in Paradise

The final report—as well as the proposals, drafts, progress reports, papers, and memos that preceded it—documents the scientists' research expertise; it also attests to their collaborative attitude toward the technical writer. As the preceding examples indicate, the researchers' ready acceptance of the technical writer as the team's communication expert is essential to professional collaboration. In an ideal collaboration, researchers rely on and appreciate the writer's superior ability to communicate their work to colleagues; further, they understand that their work remains officially nonexistent, almost as if it never happened, until it receives written documentation.

But when engineers, physicists, surgeons, or biologists do not know what the technical writer can contribute to the success of a research project, they tend to doubt the writer has much to offer. "My reports always get written, my articles published," the scientist will argue. "Besides, we've always gotten along without a technical writer before." This is a difficult argument to counter because, to a certain point, it's true. Technical writers haven't been around as long as technical writing has; to some scientists, they are still a brand new and, perhaps, passing phenomenon. It's also true that reports and articles do eventually get written. Overcoming reluctance to include a technical writer on the research team takes time; as a technical writer, you will need to be patient, tactful, and persistent.

The sometimes difficult and frustrating solution to this chronic objection lies in being ready to explain to anyone, regardless of rank, status, education, or title, what technical writers do when they write. Tell prospective collaborators the relevant points of your education and experience, particularly the subjects with which you're familiar and the kinds of technical writing you've done. Explain your familiarity with the researcher's particular discipline as well as your expertise with your own. Take time to describe who you are and what you can contribute: effective, persuasive communication of technical information to a specified audience. Don't assume that the team manager or other potential collaborators understand the concepts of "persuasive communication" or "audience"; if they did, they'd know what you have to offer the project. Some researchers are so involved in their work that they frequently overlook their readers; others never consider them. When you explain your work as a technical writer and what you can offer the research team, be sure to include your skill in audience analysis and how you can help your potential colleagues accurately meet their readers' needs and requirements. If research colleagues have difficulty distinguishing between you and their audience, ask "Do you think our readers (reviewers, editors, colleagues, grant funders, clients, students) will understand this?" Make your collaborators understand that you are asking on behalf of the audience, not because you don't understand the subject.

Although most researchers agree that technical documents need clarity, they frequently ignore the need for persuasiveness (see Chapter 3). Confident researchers can believe that the merit, purpose, or implications of their work do not need to be argued because, if written clearly, the research is good enough to stand on its own. Sometimes a persuasive demonstration of *your* ability is needed to persuade hesitant collaborators that *their* work must also be persuasively presented. For example, when I was asked to work on a grant proposal to compare the costs of terminal illness in a hospital versus the costs in a hospice, I was surprised to notice that no reasons were given for why the study was needed. The researchers wanted to examine the common but untested assumption that hospice care is less expensive than hospital care because the hospice uses family members to care for the patient and avoids hospitals' high overhead for expensive technology and life-saving at any cost. They had assumed that the purpose and worth of their proposal were self-evident. By writing a persuasive introduction that stressed the humanitarian and financial reasons for such a study, as well as its potential use by health care planners, hospital administrators, and hospital review boards, I was able to convince the reviewers that the study deserved funding. I was also able to convince the researchers that a technical writer has more to offer than a few well-placed commas.

If employers and colleagues are not well informed about what the technical writer can contribute to the research effort, they show an understandable tendency to deny the writer full collaborative status. Even when the need for a technical writer is understood, many researchers still think of the writer as an adjunct, someone necessary but not primary. This resistance to accepting the technical writer as a colleague inevitably downgrades the writer's status and responsibilities. The agricultural research company president who wanted me to meet his secretary rather than his research scientists, for example, unwittingly deprived me of professional status. In the same way, colleagues who ask a technical writer to type the final manuscript of a report or to keep an in-house library in order are not necessarily rude; they simply do not know what to expect from a technical writer.

Informing employers and colleagues of how your expertise promotes their project is the most effective way to clarify your status and identify your responsibilities. You cannot overcome an employer's or potential colleague's resistance to working with a technical writer in a single meeting, but if you have thoroughly and courteously explained your expertise as a communication specialist and how you intend to promote the team's research, most prospective colleagues will understand that you are not a member of the support staff. This process of explaining who I was, what I did, and how the researcher could benefit from my work eventually persuaded the agricultural research president to alter his original conception of how I would fit into his company.

A lack of integration into the research team not only means a lower status, it also means that the technical writer is not involved in the research project from beginning to end. But even when researchers persist in seeing the writer as part of the technical support staff, they still need their reports, articles, and proposals completed. Too many times a researcher will desperately call on the technical writer to perform verbal voodoo on a manuscript hours before it is to be mailed, presented, or delivered to the printer for final copy. Obviously this causes frustration for the writer who must attempt a rapid Band-Aid job without adequate time or information. For the researcher, two assumptions may be reaffirmed: first, that the technical writer is a normal part of the "mopping up" stage of a project, and second, since the writer's rushed job probably isn't very good, technical writers aren't really necessary.

The solution to this irritating problem, of course, requires your taking the initiative to include yourself in research projects *before* the project is due. Don't wait to be asked. Find out when meetings are scheduled and when project tests will be conducted and be there. People like to know whom they're working with, and collaboration depends in part on the technical writer's being a member of the organization's formal and informal communication networks. S/he should be part of the grapevine as well as on the mailing list for meeting schedules, administrative memoranda, calls for papers and proposals, grant and funding opportunities, and reports from other departments. The writer contributes to lunch conversations and hallway discussions as well as formal planning and review sessions. If you start when the researchers start, you will have time not only to gather all the data you will eventually need but to establish yourself as a known collaborator (Thedens, 1983).

Even after you have been accepted as a member of the research team, you may still have to counter your colleagues' doubts that you genuinely understand their scientific or technical research. The cry "You don't understand!" from fisheries biologists, R.N.'s, alcoholism researchers, statisticians, and physicists really says, "You don't have my specialized knowledge and experience, so how can you possibly claim to write this manuscript intelligently?" Almost always, however, the cry is voiced after a technical writer expresses concern over the clarity of a manuscript or its suitability for intended readers. When this is the case, you need to direct the discussion to your agreed-upon goal: a clear, effective manuscript targeted to a particular audience. Naturally, too, you will want to have as much specific knowledge of the project as you can, which you can gain through reading, observation, listening, and asking specific questions.

Technical writers, particularly those with a liberal arts education, sometimes try to alleviate their colleagues' doubts by describing themselves as generalists working with content-specific specialists. According to this argument, the generalist-writer organizes and synthesizes material but is not an expert in a particular subject. I argue that the technical writer *is* a specialist whose specialty is knowing the process required to produce effective technical communication. That is, the technical writer's specialties are skills and process specific, with a secondary but significant emphasis on content.

There will be times, even when you have been assigned to a project, that collaboration is impossible. Some researchers will refuse to acknowledge anyone but fellow researchers with their own level of education and experience. Others will insist that they've always gotten along without a technical writer and don't intend to start working with one now. If after your best efforts someone still refuses to work with you, it is probably time to seek a new assignment. Having to give up on a potential collaborator is discouraging; but remember that the collaborative failure is not yours.

References

Connors, R. J. (1982). The rise of technical writing instruction in America. *Journal of Technical Writing and Communication, 12,* 329–352.

Gould, J. R., & Losano, W. A. (1984). *Opportunities in technical communication.* Lincolnwood, IL: National Textbook Co.

Harris, J. S. (1978). On expanding the definition of technical writing. *Journal of Technical Writing and Communication, 8,* 133–138.

Polissar, L., & Eldridge, E. (1982). Cancer incidence and asbestos in drinking water in western Washington. *Washington Public Health, 3,* 21–23.

Power, R. M. (1981). Who needs a technical editor? *IEEE Transactions on Professional Communication, 3,* 139–140.

Thedens, M. (1983). Gaining the respect and confidence of the technical staff. In *Proceedings of the 30th International Technical Communication Conference* (p. W&E 70–72). Washington, DC: Society for Technical Communication.

Zook, L.M. (1983). Technical editors look at technical editing. *Technical Communication, 30,* 20–26.

10

How Can Current Computer Technology Help Technical Writers?

DEBORAH C. ANDREWS and DAVID H. SMITH

New technologies solve old problems and create new problems and opportunities. Professor Deborah C. Andrews and physicist David H. Smith provide a general overview of some problems and opportunities that arrive when computers move onto technical writers' desks. Specifically they examine two sets of problems: the complexities of available computer systems and of the market that make *choosing* a system difficult and the various barriers to mastery of a computer system. For example, the technology that makes computer systems so popular prevents them from becoming standardized and promotes incompatibility among systems. Many potential computer users perceive barriers that prevent them from trying to use a computer. Further, some users have difficulty creating a text on an electronic system either because they prefer more traditional ways of writing or because they cannot understand the often poorly written documentation that accompanies some word-processing programs.

Andrews and Smith argue that reasonable solutions to the complexity of computers center on the computer users. A computer system needs to fulfill the needs of its users the same way a functional document needs to address the concerns of its audience: one selects a computer system and software based on who will be using the system (the audience) and on how well a system will help that audience complete specific tasks (rhetorical purpose).

Despite all the assistance a computer system can offer writers, it cannot substitute for a writer. It will help a writer make a document look better and, perhaps, read better. However, it will not make a poor writer into a good writer. The technical writer is still the communications expert who uses technology's newest tool, the computer.

Computers as Solutions and Problems

Sometime in 1984, capital investment per office worker in America for the first time exceeded that per factory hand. Much of that investment created computing power to produce and manage information and send it through electronic networks that span rooms, offices, cities, and even the globe. The potential for such technology is vast. But at present, although we talk about *computers* and *computing* as if they are one thing, the technology is diverse, much in flux, unstandardized, and not integrated. Prices fluctuate widely. The system that you choose today may not be compatible with another you find you need in a month. Selecting the *right* system is difficult. That's problem 1.

Second, new technology presents a set of barriers to one's mastery of it. One barrier is psychological. Some people develop knots in their stomachs when they face a terminal and must be cajoled into trying out a system. Some people have a hard time conceiving of a document that isn't on paper. They don't trust the computer and need paper to know they've written prose.

Another barrier is manual, that is, a lack of the necessary keyboarding skills. Another is cultural. At large technical organizations in particular, some managers lock engineers and other professionals out of word-processing systems because it seems inappropriate behavior for a professional to type.

Sometimes, too, the documentation that accompanies a system is a barrier to using the system. That picture may be changing as technical writers take charge of documentation and vendors realize that the documentation, almost more than the hardware, sells a system. Still, many manuals hinder more than they help.

The computer's behavior, too, may be a barrier. A computer is unforgiving. If you make a mistake, it counts, sometimes with dire results. Even ardent users approach terminals in a somewhat adversarial spirit and with a certain amount of gamesmanship. And a computer makes demands. You have to reshape your thinking about routine writing tasks to frame what you need and your instructions to the computer in a code it can read. In general, computers reward those who work in an orderly manner. They also require some housekeeping—more housekeeping than a pencil or a typewriter requires.

Some or all of these barriers may apply to you. Your writing has to wait while you master the system. Simple routines can be learned in a few hours. But getting comfortable with more powerful techniques may take weeks. Such learning is hindered by the pressure of deadlines that may cause you to revert to a familiar routine. These barriers, then, form problem 2.

Selecting a Computer System

Having laid out the two big problems, let's look at some approaches to the solutions. First, selection: any decision on computing, including word processing, should start with the user, not with the equipment. To begin, develop a profile of your work, or your group's work, and your own composing style. That's the baseline. Here are some questions to think about:

- What kinds of documents do you write?
- Do you need to hold and manipulate different documents that merge in and out of each other?
- How much time can you devote to learning the system?
- Do you need extensive formatting capabilities?
- Do you frequently write in groups and need to circulate documents electronically?
- How many people will share the system?
- How many documents must you store at a time? How frequently do you access those documents?
- Do you require extensive footnotes?
- Do you work for a multinational corporation that documents in languages?
- Must the system be compatible with a graphics package or statistical package or other computer-aided function (like Computer-Aided Design/Computer-Aided Manufacturing) used in your company?

Think about what you want the computer to *do* for you. Then choose the software, the programs that can relieve you of some of this work.

Programs

Some programs emphasize *output*. That is, they provide versatility, polish, and reproducibility—even thousands of copies—in final publication. The most sophisticated of these programs turn your desk into a print shop. They require powerful computers, usually the shared mainframe systems available to large organizations.

Other programs emphasize *input*. That is, they make it easy for writers to write, to enter text. Many such programs run on microcomputers, the small, sometimes portable machines that serve a single person, or on dedicated word processors, equipment used just for this purpose.

Of course, as a writer you need to get text both *in* and *out*. But publications managers may emphasize final publication forms over a writer's needs. Individual writers may want a system that coddles them. The term *word processing* covers a variety of ways computers and people work together to produce text. One simple program may combine many functions for entering, editing, and printing text. But on larger computers the functions of word processing may be segmented. You may use a *text editor* to write, edit, and store the text and another program, called often a *word processor,* to impose automatically preferred page formats and printing characteristics.

In determining ease of use you may compare several measures. One is the total number of editing commands the program offers. Another measure is the speed and ease with which you can move through the text and locate particular items for change. A third measure is the number of keystrokes required to make each change; here, of course, the lowest score is best. In addition, most users prefer commands that are easy to remember. Cursor keys with arrows, for example, are easier to use in locating text than letter designations for movement (*k*, for example, to move left in the text; *h* to move right, etc.). For extensive editing, many technical writers prefer what-you-see-is-what-you-get programs that show you on the screen exactly what your text will look like as it's printed. Some programs require the placement of codes for margins, format, etc. on the screen and only show the final layout in print. Such designations may increase difficulty in editing and proofreading.

To find out what's available, you will want to consult the literature on computers and computing, a literature that multiplies daily.[1] Much of the literature is descriptive and evaluative of new equipment and programs. Some researchers, however, are beginning to examine links between the technology and productivity. Computers also provide an interesting tool for analyzing writers' composing processes, as Bridwell (1983), Daiute (1983), and others have shown. But at this point the technology is still new enough that researchers are cautious about drawing conclusions. Much of the evidence is still only anecdotal—useful as that may be. Most preliminary results point toward the benefits of computing, but the euphoria may be tempered in the longer run. For now, use the literature mainly to narrow the number of systems you'd like to test drive.

As you read and try programs you might think about how your work profile fits into these three categories:

- *Pencil and eraser* programs aimed at ease of use.
- *Assistant and editor* programs that automate several writing and editing functions.
- *Print shop* programs that emphasize polish in output.

Pencil and Eraser

As a writer you may simply require an electronic pad, pencil, and eraser. You want a program you can master in a few minutes to create short documents without elaborate format demands. Programs that use graphics for instructions (like a trash can for items to be deleted) and a mouse (a little hand-held device) for locating items rapidly on the screen accommodate this purpose well. So do programs developed for children, particularly those in elementary schools.

Assistants and Editors

But with some packages you trade off ease of use against access to the full processing capabilities of a computer. You may be willing to invest more of your own learning time (and usually more money) for more powerful systems.

At one level, good word-processing software serves as your assistant in preparing automatically such standard document elements as tables of contents, title pages, indexes, footnotes, and bibliographies. It can paginate, create headers and footers (text repeated on every page), and count the total number of words you use. The software can also allow you to take notes for eventual retrieval in a text, merge texts from several documents, set up tables automatically, and otherwise encode research results.

At another level, the software serves as your copy editor. Several programs check spelling throughout a text, matching your document to electronic dictionaries of hundreds of thousands of words stored in the computer's memory or on a disk (although such programs can't detect homonyms or commonly confused terms like "effect" and "affect"). Many programs also include "search and replace" features that correct automatically items entered incorrectly at several places in the text.

At the highest level, some programs act as your alter ego and electronic reader with advice on grammar and style. Probably the most sophisticated and best known of these is the Writer's Workbench program developed by Bell Laboratories and marketed by AT&T Technologies (see, for example, Cherry, 1982; Macdonald et al., 1982). This package of programs and databases both proofreads and advises. It checks text for spelling, punctuation (like unclosed pairs of parentheses), double words, certain awkward phrases, and sexist language. Its textual analysis programs can measure your prose for readability against either standard measures (like the Kincaid readability scale) or ones you program in yourself. It can also print out all your headings and all the opening and closing sentences of paragraphs to give you a check on cohesion. It will assemble a list of your most frequently used nouns and adjective–noun pairs and can search for acronyms. It can recognize parts of speech and count most anything you might want to count in your writing.

IBM has a similar program, called "Epistle," which now runs on large computers but will probably be available for personal computers as well. It can note parts of speech and dependent clauses and is being programmed with aids to style.

Such programs can remind writers about deviations from standard practices. If your style is confused or ponderous, and you're aware of this, the programs may prod you to reform. But because they are somewhat time-consuming to run and arbitrary, they may be of only limited use to seasoned writers.

Print Shops

Finally, some software, primarily programs that run on mainframe computers, can turn an office into a print shop. One large aerospace manufacturer, for example, has developed an automated production system that runs on a minicomputer and will handle the division's total output of manuals—some 76,000 pages a year. The system includes several terminals, at which operators enter text, and two terminals for graphics. In the old system pages had to be pasted up by hand from long rolls of galleys. Now, the computer formats text for each page according to guidelines the supervisor sets in the beginning. The guidelines can be varied for each new document. Variations can also be tested easily; operators can enter codes for different margins and typeface to see how the text will look and can obtain quickly full-page layouts for each possible design. The program calculates room for graphics and can print the graphics too, if desired. The system also accommodates many typefaces and changes in headings automatically.

Such formatting capabilities of computers—the ability to lay out a page and vary features like type fonts, pitch, and print characteristics—are particularly exciting to technical writers and publications managers. Major mainframe systems developed specifically for text production are extraordinarily versatile and can create in-house documents that previously were within the range of only the most sophisticated typesetters and printers. The publications manager thus increases control over output and speeds up production, usually with a long-term financial savings.

The final polish of the text results from the type of system you use and your choice concerning the polish necessary to do the job. At one end, dot matrix printers are relatively cheap and fast. The resolution may not be high, but the text is readable and is often exactly what you need for drafts and sometimes for final documents. Line printers give similar resolution for text from mainframe computers. Better resolution print comes from letter-quality printers that are, however, usually more expensive than dot matrix ones and slower. At the other end are phototypesetting, laser, and ink-jet systems available to institutional users. These match the quality of professionally published text. Any one document you are working on can be sent through different stages of printing to accommodate whatever degree of polish in the form of the document you desire. The need to compromise on polish is reduced when a computer enters the picture.

Mastering the System

If you have selected well, installing and learning the system should not be difficult. To overcome the barriers we cited, keep chiefly in mind the need to work in increments and with courage. Start with the easy procedures and work toward the more difficult ones. Don't try to do everything at once or let yourself be frustrated by the capabilities you are *not* using. Perhaps you don't need them. At least, you can wait awhile to work up to them. Many companies appoint certain staff members as troubleshooters to help individual writers learn the system. Users groups can also ease the transition into knowledge through meetings to share problems and solutions. Managers should understand that not all typing is secretarial; increasing numbers of technical people are learning to *think* at a keyboard, to capture ideas quickly on their own rather than depending on others to transcribe them imperfectly.

The opportunities that come with mastery—even limited mastery—are probably familiar. But let's check off a few of the most significant benefits and then provide some cautions.

Speed
First, *speed*. You can write as fast as you can type, which is probably faster than you can write longhand. Writing on a computer is like dictating but with the advantage that the product is under your own control and doesn't depend on someone else for transcription.

Ease of Revision
Second, *ease of revision*. In part, you can be speedy in drafting because it's easy to change the text. Composing at the screen removes the mess of changing words around. The new text instantly replaces the old. You can make little or big changes with equal ease.[2]

If you like to share a text with colleagues as part of your revising process, the computer makes that easier too. For one thing, handwriting can be hard to read, both in a longhand original and in jottings on an otherwise typed manuscript. But even a quickly composed draft is readable when it's printed from a computer. Moreover, on a mainframe and increasingly with the networking of microcomputers, it's easy to send text around to coauthors and editors. No paper need change hands. And because readers know you can incorporate change easily, they are often more willing to comment on substantive issues and offer suggestions.

Be cautious in using a computer to revise, however. Because of its very ease and because of a certain hypnotic quality to a computer screen, revision there may be pathological. You can find yourself never letting a sentence sit still. You may exhaust your energies on frivolous details and rearrangements to the exclusion of real substantive reexamination. You may also absorb energy in computer housekeeping, playing with files and commands rather than text. Such exercise has two drawbacks. First, your text suffers because your attention is elsewhere. Second, if you're new to a system, you may lose text in entering commands you don't understand. Many systems provide safeguards against just such accidents—the system may ask you, for example, if you *really want* to delete something—but the possibility of loss is real. Moreover, unless you specifically program your computer to save all changes, or unless you have access through machine language to the inner workings of your terminal, you will lose any text you replace with other text. Most of the time that's probably fine. But it is one of the costs. Our final word on something is not always our best; the others, however, are not practically preserved on the terminal.

Integration
A third benefit is *integration*. By "integration" we mean the many ways that word processing can be linked to other computer systems for information processing. Although the technology is still in a fledgling state, integrated packages are a goal of leading software manufacturers. A database management program may contain an electronic spreadsheet, a graphics package, a system for developing outlines from notes, even a tie-in to systems for managing production, all easily merged into a word-processing program for document preparation and printing. Even now, many programs allow for easy preparation of form letters tailored with personal elements and circulated either through conventional mail or electronically.

Finishing
Fourth, *finishing*. We've noted the print-shop polish increasingly available with even simple computer systems. Proofreading programs also reduce drudgery and (one hopes) increase correctness in prose. Beyond correctness, too, computers can enhance readability by augmenting the signals writers can give to readers through varied typefaces and boldfaced for

headings and text elements to be highlighted. Such *macropunctuation* provides markers for segments of a text in the way that punctuation within sentences groups and separates phrases and clauses. The computer didn't create the possibilities for these markers, but it does foster such format devices.

And because letter-perfect copy of the final draft can be prepared in minutes on a printer, you can hold a document in suspension longer before the due date, incubating new ideas and new language until shortly before you submit the document. Reports thus gain in both thoroughness of time and timeliness.

Relevant Research: Keeping out of Trouble

These benefits, of course, cost. Let's look at a few cautions here for keeping out of trouble. Again, these may be familiar to you.

Save and Backup Text

Most writers who have shifted to computers can tell horror stories about losing text. A writer, daydreaming in a document, carelessly runs a finger across the keys and pushes, by accident, the reset button. Whosh!—gone in a flash is two hours' work. A writer on a mainframe system marches along the line and then sees on the screen a jumble of letters. The system has crashed. Gone is the text.

The fear of loss, however, is easily assuaged. You simply have to get into the habit of saving the text at frequent intervals—say, every page, or every 15 minutes. Set a timer to remind you. Or save text whenever you have filled a screen or see a new page indicator. Save text *before* you answer the phone. At the end of the day or a writing session, backup the whole file. Make a copy of the file on another disk and store that disk in a different location from the disk with the original. Or copy the file for storage on a mainframe. The really paranoid make three copies: two on disks and one in print.

Organize Files

Computers reward the orderly. Take time to organize files—even though disorder isn't as apparent on a disk as in a file cabinet. Name documents with richly retrievable terms (not just "Proposal") and perhaps keep a list of which documents are on which disks. Attach a note card to the envelope of the disk with the file name and a brief descriptor. The word-processing system you are all using will provide you with a directory of files on the system, but this directory is displayed in the computer's order rather than yours and may be hard to skim.

Select file names, then, carefully, indicating both the *content* and the *version* of the draft. You need to note the version because you may, in backing up text, store different drafts. You want to make sure you print out and circulate the right one. Here, for example, is one horror story in support of this need. Three students, working together on a manual, kept private records of versions of text and a general record of responsibility for each chapter. One version of a chapter included a coding error that was corrected in the final version. But another student, compiling the whole manual at the end of the term, printed the wrong version. The result was 1,300 pages of paper spun off a high-speed printer bearing only the header on each page—and no text.

Coding blocks of prose—text to be repeated in different documents—poses even greater problems than creating file names for single documents. You might use both numeric and verbal codes: numbers for originating department, a letter for the type of document (proposal, for example), another letter or series of letters describing the specific content.

Finally, limit file size. The page limit of a paper file is roughly the amount you can fit into a folder. A computer file is limited by the amount of memory in the system. That may be eight pages on a small computer or 50ish on a floppy disk or thousands on a mainframe. Different systems treat the storage and recovery of files differently. Usually, however, the amount of time it takes to search through and save a file is directly proportional to its length. Thus, you may be discouraged from frequent saving of a file if it's long. Practically, it's best to limit file size to around 20 to 30 pages, particularly on microcomputers. The files are thus more manageable.

Store Disks Carefully

The box that contains the floppy disks for your microcomputer probably is covered with various warnings. Adhere to these. Watch out for coffee and scratchy pens that might rub out your text. On the other hand, you needn't feel that a computer commits you to a sterile room for writing. You can still sip coffee near a computer and your text will live to tell the tale. Just sip carefully and don't spill on the disk.

Know Your Limits

Some writers complain of eyestrain at the terminal if they invest more than a few hours at a time. Backs, too, may give out. Early in your use of the computer, if you experience eyestrain or backaches, correct for any excesses.

Conclusion

In spite of the claims of some manufacturers and some of the more boosterish magazines, a computer is not going to turn a poor writer into a great one. It will also not do all your work for you. And as it aids you it also requires other sorts of work in tending its needs and increased numbers of decisions about the documents you are producing.

In the end, you remain a writer first and a computer user second. It's simply one mode of writing to be exploited, along with pencils and pens, whenever it serves you. The likelihood is great, however, that technical writers, journalists, creative writers—people who cherish prose—will derive great benefit from the new technology. A computer is yet another tool to ease your lot in refining prose, to enhance the possibilities for printing that prose in documents that both clarify the message and please the reader's eye, and to foster enjoyment while you work.

Notes

1. Indeed, one difficulty in choosing and using computer systems is the profusion of literature on the subject, much of it contradictory, and some of it in pretty fugitive sources. The brief list of "Suggested Readings" at the end of this chapter provides some recommended source materials. In general, you are likely to find good articles on the topic in the *Proceedings* of the International Technical Communication Conferences held each year by the Society for Technical Communication and in the society's journal, *Technical Communication;* in the *IEEE Transactions on Professional Communication;* in the *Journal of Technical Writing and Communication;* and in *P.C. Week, Byte,* and *Personal Computing.*

2. To aid in your revision at the screen keep a notepad by the computer's side to jot down locators in the text. You can only see a screen—some 14 or 24 lines, depending on the system—at a time. That's fine for drafting and for close editing sentence by sentence. But it does limit your ability to skim a text for large-scale revisions. Most systems provide strategies for moving page by page or rapidly to the beginning or end. But you still may have problems remembering the particular form of headings or the section you are in as you dive farther and farther into the details. Write down each new heading on the notepad as you put it on the screen. Note the page and line of the heading (these details are usually displayed at the bottom of the screen). In that way you can see the frame of the text and find sections easily. If you are working from an outline, of course, you can just add locator information to each topic on the outline. Some systems can create a "Table of Contents" from your headings automatically. You may be able to call this up

as you write, in a window or on the full screen, and work from there rather than a notepad.

References

Bridwell, L. (1983). *Computers and composing: Implications for instruction from studying experienced writers.* Paper presented at the meeting of the Conference on College Composition and Communication, New York.

Cherry, L. (1981). Computer aids for writers. *Proceedings of the ACM SIGPLAN,* **16–6,** 61–67.

Cherry, L. (1982). Writing tools. *IEEE Transactions on Communication,* **30,** 1, 100–105.

Coke, E. A. (1982). Computer aids for writing text. In D. Jonassen (Ed.), *The technology of text: Principles for structuring, designing, and displaying text,* Vol. 1 (pp. 383–399). Englewood Cliffs, NJ: Educational Technology Publications.

Collier, R. (1983). The word processor and revision strategies. *College Composition and Communication,* **34**(May), 149–155.

Daiute, C. (1983). The computer as stylus and audience. *College Composition and Communication* **34** (May), 134–145.

Fluegelman, A., & Hewes, J. (1983). *Writing in the computer age: Word processing skills and style for every writer.* New York: Anchor Press/Doubleday.

Glatzer, H. (1981). *Introduction to word processing.* Berkeley, CA: Sybex.

Howard, J. (1982). Advances in computer technology: What will the impact be for the professional communicator? In *Proceedings of the 29th International Technical Communication Conference* (p. T-28-30). Washington, DC: Society for Technical Communication.

Macdonald, N., Frase, L., Gingrich, P., & Keenan, S. (1982). The writer's workbench: Computer aids for text analysis. *IEEE Transactions on Communication,* Com-30, **1,** 105–110.

McWilliams, P. (1983). *The word processing book: A short course in computer literacy.* Los Angeles, CA: Prelude.

Nancarrow, P., Ross, D., & Bridwell, L. (1984). *Word processors and the writing process: An annotated bibliography.* Westport, CT: Greenwood Press.

Rothman, M. (1980). The writer's craft transformed: Wordprocessing. *Oncomputing,* **2,** 60–62.

Schwartz, H. (1982). Monsters and mentors: Computer applications for humanistic education. *College English,* **44,** 483–490.

Zinsser, W. (1983). *Writing with a word processor.* New York: Harper & Row.

11

How Can Technical Writers Give Effective Oral Presentations?

PETER WHITE

Professor Peter White argues that oral presentations produce more anxiety than the production of a written document. When students, engineers, doctors, scientists, and even experienced teachers are called upon to speak to colleagues in a formal environment, the results can vary from entertaining and informative to boring and hopelessly confused. All professionals realize that public presentations constitute a major investment in time, energy, and money.

Whether it's called a speech, a report, a briefing, or a professional paper, the oral presentation of information demands that the speaker enter into tacit, general contracts with his profession at large and local, more specific contracts with the audience and organizers of the specific speech event. These implicit contractual agreements ask the speaker to make detailed preparations that will take into account a frank assessment of the nature and value of his information as well as the best means for delivery of that information.

White also argues that the first step in the execution of these contracts is a sound understanding of the nature of the event. To explain theoretical aspects of public performance, White relies upon the work of sociolinguists Erving Goffman and Kenneth Pike, figures not generally known to most technical writers. Both Goffman and Pike see the lecture as a conventionalized, dramatic, or ceremonial construct: it is an artificial exercise, a spectacle and game, a "frame" of experience. In this regard, speakers are animators, authors, and principals who may choose to deliver their messages in varying combinations of memorized speech and fresh talk, speaking that gives the illusion of spontaneity. The speakers attempt to minimize the possibility of "production crises" and other offenses against effective interaction while, at the same time, they struggle to meet the major goal of presenting a clear delineation or portrait of some meaningful part of the objective world. In

191

other words, out of illusion and play speakers propose a description of significant reality.

Speaking Persuasively

Although Woody Allen is somewhat reassuring in his belief that 90% of life consists in merely showing up, professional writers and speakers are more concerned with the remaining 10% of life when they must *do* something; they must perform. Federal Express realized this problem in a recent advertising campaign when they portrayed a junior executive reduced to making hand shadows at a corporate board meeting because *he* had arrived for the presentation but his slides had *not.* The commercial is funny, but the reality it plays upon is serious. To restate the drama of this commercial in more academic terms: in the professional world this junior executive plays the role of *Everyman,* alone at the lectern, placed in a testing environment, scrutinized by the powers to whom he appeals, but finally helpless and ridiculous—unable to perform, persuade, or act.

Like facing an important job interview or acting in a play, giving a speech demands that individuals present themselves in a performance situation. Speakers must intellectually expose themselves in a context where their thinking is open for inspection, where their thoughts are tested and evaluated, and where there seems to be potential for public failure and even a form of humiliation. To draw upon another television advertisement—this time for American Express—the real question for technical communicators faced with the challenge of giving an oral presentation is simply put: "What will you do, what will you do?"

Relevant Research

Primary research in the field of technical communication indicates that engineers, scientists, technical personnel, and managers are acutely aware of the pressing need to learn the art of giving better oral presentations. *Engineering Education* surveyed 367 engineers in chemical, civil, electrical, geological, mechanical, metallurgical, and mining engineering and asked them to rate 30 specific communication tasks for job-related importance. The respondents rated 24 tasks as important, 14 requiring writing skills, and 10 oral skills. Of the 10 oral tasks, 5 involved giving formal oral presentations, including project proposal presentations, illustrated presentations (those using charts, graphs, slides, and other aids), progress reports, feasibility studies, and formal speeches to sophisticated colleagues (Kimel & Monsees, 1979). Andrews and Blickle (1982, p. 382) report that

> One company, General Electric, notes that only about 25 percent of their engineers' time is spent in actually doing engineering activities; 75 percent is spent in communicating before and after (and sometimes interrupting) the doing. And much of that time, for engineers and other professionals, is spent on the phone, in meetings, and in formal and informal presentations.

Geonetta (1981) reports that over 3,500 engineers responded to a questionnaire aimed at identifying the areas in which they needed intensive training. Sixty percent of the respondents selected public speaking as the fourth most important area. In Pennsylvania approximately 75% of 2,000 engineers surveyed said they needed more training in basic communication skills, including public speaking and conference leadership. One author reported that business executives spend almost 70% of their working hours in conferences where oral communication is the predominant mode of sharing information and ideas.

Engineers and executives have come to realize the importance of public speaking through simple calculations that reveal the financial waste involved in an unsuccessful presentation. By updating Schoen's (1979) salary statistics for the mid-1980s, we can see that a 20-minute talk attended by 200 professionals paid an average of $40,000 will cost a minimum of $1,280. Research and development corporations have only the written and the spoken word to represent their "product," and people will not buy an idea they cannot understand.

There are several ways of looking at what constitutes the *problem* of giving a formal presentation. On the one hand, we can view the scientific and technical community as a body of individuals sincerely interested in communicating discoveries, methods, recent developments, or the latest technology to peers, customers, or the general public. Somewhat less idealistically, however, we can describe conferences, presentations, seminars, and other professional gatherings as merely occasions for compulsory attendance, crass salesmanship, diversionary entertainment, and, of course, pomposity and egotism. Therefore, what technical writers or scientific experts actually *do* in a formal presentation, and in fact what is received by the audience at such an event, are conditioned by the attitudes of both the speakers and the listeners toward the public or communal components of their profession.

What speakers *will do*, then, depends upon how they define the nature and goals of public communication. The first and greatest problem for public speakers is establishing the *general contract*, the agreement speakers make with themselves about the value of delivering information or findings to others, including their real intentions, their motives for undertaking the task, their conscious willingness—indeed, their desire—to be complete and clear, and finally their candid assessment of the existence of certain human factors (insecurity, egotism, ambition, among others) that frankly compete with the more altruistic aims of science, business, and academia. To form the general contract speakers need to identify and evaluate the conflicting thoughts and drives that presuppose and shape their performances. Although it may seem natural and obvious that professionals need to know the terms of their general contract, there is remarkable evidence that for the audience such matters have usually not been settled.

Houp and Pearsall (1980) presented the thoughts of an "educated" audience as they listened to an "excellent" speaker at the National Training Laboratory in Group Development. The following passages represent selections from the "introspections" of these auditors, jotted down during the speaker's presentation.

> God, I'd hate to be speaking to this group. . . . I like Ben—he has the courage to pick up after the comments. . . . Did the experiment backfire a bit? Ben seems unsettled by the introspective report. . . . I see Ben as one of us because he is under the same judgement. . . . He folds his hands as if he was about to pray. . . . What's he got in his pocket he keeps wriggling around. . . . I get the feeling Ben is playing a role. . . . It is interesting to hear the words that are emphasized. . . . This is a hard spot for a speaker. He really must believe in this research. . . .
>
> Ben uses the word "para-social." I don't know what that means. Maybe I should have copied the diagram on the board. . . . Do not get the points clearly . . . cannot interrupt . . . feel mad . . . More words. . . . I'm sick of pedagogical and sociological terms. . . . Slightly annoyed by pipe smoke in my face. . . . An umbrella dropped. . . . I hear a funny rumbling noise. . . . I wish I had a drink. . . . Wish I could quit yawning. . . . Don't know whether I can put up with these hard seats for another week and a half or not. . . . My head itches. . . . My feet are cold. I could put on my shoes, but they are so heavy. . . . My feet itch. . . . I have a piece of coconut in my teeth. . . . My eyes are tired. If I close them, the speaker will think I'm asleep. . . . I feel no urge to smoke. I must be interested. . . .
>
> Backside hurts. . . . I'm lost because I'm introspecting. . . . The conflict between introspection and listening is killing me. Wish I didn't take a set so easily. . . . If he really wants me to introspect, he must realize himself he is wasting his time lecturing. . . . This is better than the two hour wrestling match this afternoon. . . . This is the worst planned, worst directed, worst informed meeting I have ever attended. . . . I feel confirmation, so far, in my feelings that lectures are only 5% or less effective. . . . I hadn't thought much about coming to this meeting, but now that I am here it is going to be O.K. . . . Don't know why I am here. . . . I wish I had gone to the circus. . . . Wish I could have this time for work I should be doing. . . . Why doesn't he shut up and let us react. . . . The end of the speech. Now he is making sense. . . . It's more than 30 seconds now. He should stop. Wish he'd stop. Way

over time. Shut up. . . . He's over. What will happen now? . . . (Houp &
Pearsall, 1980, pp. 427–428)

Houp and Pearsall originally used these excerpts to impress upon technical writing students the need to deal with audience inattention. But such
"introspections" reveal much deeper and more persistent problems: the
speaker is troubled by the pressure; attacked for role playing; ridiculed for
obfuscation, distracting mannerisms, and reliance upon jargon; portrayed
as a tyrant who refuses to interact; and is even condemned as an intermin-
able bore. Through their own words, the members of the audience display
childish traits: fear of being placed in the limelight, guilt over not taking
notes or paying attention, resentment of the physical restraints, anger directed at the authority figures and organizers, and longing for freedom and
escape. This is a disturbing but realistic picture of the event professionals
call the oral presentation. The communication is not working—the speaker
is not effective—because few seem to have a clear understanding of the
purpose and the value of this speech.

Having settled upon the terms of the general contract, the speakers' second problem is to arrive at a specific definition of the localized event. Who
will attend the speech and why? Are the speakers expected to entertain,
sell, teach, persuade, move others to act or merely inform them? How well
informed will the audience be, and upon whose behalf will the information
be imparted? What are the exact restraints placed upon speakers, including
time limits, access to audio–visual equipment, and the environment of the
meeting? This more *local contract* requires that speakers address themselves to the actual audience assembled before them, to the people who
have come to hear a specific message that has been tailored to meet their
professional needs. Forming the local contract means much more than
dragging out bits and pieces from previously delivered lectures. Truly adequate preparation involves a complete restructuring of the content and
the organization of the material, a thorough examination of the best possible modes for delivery, sometimes major adjustments in the types of audio–
visual aids, and so forth. To give a successful presentation speakers have
to use commonsense psychology and do a good deal of homework. They
should spend time talking with the organizers of the event and ask for any
written materials that might help them to analyze the audience, their professional level of achievement, their schedule before and after the event,
their institutional affiliations, and any problems that they may share as a
group, including their common biases and prejudices. Speakers must realize that they are addressing people—scholars, scientists, lawyers, clerks,
mechanics, technicians, doctors—and not a subject; priority should be given
to the audience.

A professional talk is actually a dynamic event involving many different factors and perspectives. Summarizing years of experience at conferences and professional meetings, White (1979) listed the thoughts that go through the minds of speakers and listeners in the international scientific community. The speakers wonder:

- Will this material stand the scrutiny of my peers?
- Does it advance the state of the art as they perceive it?
- Will someone in front row ask the evident question I've overlooked which obviates the basis of all my work?
- Will I be able to tell the whole story in 15 minutes?
- How will I fill up 15 minutes, all the conclusions being obvious to me in five minutes?

The listeners, on the other hand, ask:

- Why was this work undertaken in the first place?
- What motivates this line of attack?
- What are the physical principles behind the method of solution or the area of study?
- Why are we proceeding through this elaborate derivation in the talk?
- Why can't the equations be listed in a digest and a qualitative explanation be given in the talk?
- What ideas and techniques can I take from this talk which will help me in my work? (White, 1979, p. 179)

White shows the speakers and the listeners on two different emotional and intellectual tracts. Speakers are most concerned with their own image, with their professional competence, or with avoiding huge embarrassment, and with thinning or stretching the material. The listeners are pragmatic: they want to be lead logically and clearly through the material to the main points and principles so that they can profit from this new knowledge. They want to be taught so that they can advance in their own work. The important general conclusion is that both parties are self-centered in the dynamics of the presentation: speakers are anxious about their *person,* and the audience is present to reap some *benefit.* Beneath the public talk—just below the surface—there is conflict, and perhaps this is the problem of the oral presentation, a dynamic give and take between parties who have something to lose and gain.

Theory

The most enlightening work on public speech has been done by linguist Erving Goffman (1959, 1974, 1981). Although his work is not well known among practicing technical writers/communicators, he can be useful in helping us to understand the social anthropology of the public performance. Goffman is fascinating because he looks at our language and our behavior as an ethnologist studying some exotic tribe, unraveling the ways in which individuals present themselves, as actors on the stage, to the audience we call society at large. As Goffman succinctly states in the preface to *Forms of Talk,* "I make no large literary claim that social life is but a stage, only a small technical one: that deeply incorporated into the nature of talk are the fundamental requirements of theatricality" (Goffman, 1981, p. 191). Goffman (1981, p. 9) further explains:

> Now let me take another try at saying what it is that a speaker brings to the podium. Of course, there is his text. But whatever the intrinsic merit of the text, this would be available to readers of a printed version—as would the reputation of its author. What a lecturer brings to hearers in addition to all this is added access to himself and a commitment to the particular occasion at hand. He exposes himself to the audience. He addresses the occasion. In both ways he gives himself up to the situation. And this ritual work is done under the cover of conveying his text. No one need feel that ritual has become an end in itself. As the manifest content of a dream allows a latent meaning to be tolerated, so the transmission of a text allows for the ritual of performance.

A "speech event" is part game, part spectacle, as basic or ceremonial as any tribal expression, in which an individual is tested through a manufactured and conventionalized setting. Goffman presents a striking analysis of the thought processes of speakers, the various forms of language appropriate to the event, the interaction of speakers and audiences, the requirements of setting, and the expectations held by the organizers and by all present. At the center of all this dramatic and dynamic activity, Goffman (1981, p. 173) places what he calls the "textual self, that is, the sense of the person that seems to stand behind the textual statements made and which incidently gives these statements authority. Typically, this is the self of 'relatively long standing.' " In a sense, Goffman is saying no more than Aristotle proposed in his *Rhetoric,* that effective persuasion depends upon the credibility or moral character of the speaker.

Briefly, Goffman argues that auditors have access to the ritualized projection of the speaker's self in ways that can never be completely duplicated through the written word. Thus, having this direct accessibility to authority, "the audience also gains ritual access to the subject matter over which the speaker has command" (Goffman, 1981, p. 187). Of course, very few speakers are consciously aware of the dynamics here; it is the scientists' or the technical communicators' intuitive sense of the situation that causes anxiety they cannot properly and positively channel because they do not fully understand the true source of the discomfort. They do not actively understand that they have contracted with their audience to perform a role as leading actor in a highly formularized expression of human behavior. They do not feel sure of the context, the roles of fitness, propriety, aptness, and decorum for a particular speech event.

Following the lead of linguist Kenneth Pike (1967), Goffman (1981, p. 167) calls the lecture "a laminated affair of spectacle and game" in which the speaker is *animator* (the voice machine), *author* (the mind behind and writer of the ideas), and *principal* (the authority who believes in the scripted position). Speakers choose one of three basic forms of communication: *memorization, aloud talk,* and *fresh talk.* Each of these "production modes" presupposes a different relationship between speaker and audience, "establishing the speaker on a different 'footing' in regard to the audience" (Goffman, 1981, pp. 171–173). *Memorization,* the most formal and theatrical is studiously avoided, whereas most speakers engage in "production shifts" from *aloud talk* (reading) to *fresh talk* (extemporaneous composition). Fresh talk is the most desirable but also the most illusory. To understand how Goffman penetrates the mental processes of public speaking, consider his explanation of the nature of fresh talk and his description of facility in this production mode:

> It might be noted that fresh talk itself is something of an illusion itself, never being as fresh as it seems. Apparently we construct our utterances out of phrase- and clause-length segments, each of which is in some sense formulated mentally and then recited. Whilst delivering one such segment one must be on the way to formulating the next mentally, and the segments must be patched together without exceeding acceptable limits for pauses, restarts, repetitions, redirections, and other linguistically detectable faults. Lecturers mark a natural turning point in the acquisition of fresh-talk competence when they feel they can come close to finishing a segment without knowing yet what in the world the next will be, and yet be confident of being able to come up with (and on time) something that is grammatically and thematically acceptable, and all this without making it evident that a production crisis has been going on. And they mark a natural turning point in fresh talking *or* aloud reading a lecture when they realize they can give thought to how they seem to be doing, where they stand in terms of finishing too soon or too late, and what they plan to do after the talk—without these backstage considerations becoming evident as their concern; for should such preoccupation become evident, the illusion that they are properly involved in communicating will be threatened. (Goffman, 1981, p. 172)

Anyone who has made a career of speaking in public on a regular basis will instantly recognize Goffman's insight: good lecturers are capable of achieving a sort of dual consciousness, being at once the "actors" and the "audience" able to "see" themselves performing while they create the dramatic monologue or illusion that in fact becomes the "play." Through experience and inner self-confidence, they are comfortable and relaxed as they make mental formulations; repair syntactical, thematic, logical, and delivery faults; and move through the areas that Goffman calls "production crises."

Folklorist Roger D. Abrahams, commenting on Goffman's theories, would

describe such notions of the "production crisis" as equivalent to philosophical alienation (Abrahams, 1984). When Goffman employs the term
alienation, says Abrahams (1984, p. 83),

> it is to show what preparations we make for the possibility of embarrassment
> and other "offenses" against the flow of talk. It is almost as if he regards
> faulty interactionists as sinners—or at least failed humans—for as humans we
> must be willing to give ourselves up to moments of engagement . . . for such
> "moments are of transcendent importance."

Goffman takes "offenses" against interaction so seriously because the lecture is a "frame," that is, an order of experience that is set off from the
mundane, that is intense and highly focused, having boundaries of time
and place (rules) that must be followed so that both the lecture and society
at large have authority, foundation, and ultimately real meaning. In a public address, speakers are the crucial links or "brokers" between the audience and the information or "message" it desires to have.

For technical communicators, then, the oral presentation carries tremendous responsibilities, obligations that in Goffman's words resemble those
of science itself. Goffman (1981, pp. 194–195) states that the lecturer and
the audience

> join in affirming that organized talking can reflect, express, delineate, por-
> tray—if not come to grips with—the real world, and that, finally, there is a
> real, structured, somewhat unitary world out there to comprehend. . . . And
> here, surely, we have the lecturer's real contract . . . to stand up and seriously
> project the assumption that through lecturing a meaningful picture of some
> part of the world can be conveyed, and that the talker can have access to a
> picture worth conveying.

Ledger (1982, p. 39) objects to Goffman's ideas. "You do not feel that
you can *do* anything with what you learn in 'The Lecture,' except to be
acutely aware of what a lecturer does. . . ." Furthermore Ledger (1982,
p. 39) feels that Goffman's deeper point resides in his movement "beyond
sociology and its implicit ties to practical application to a sort of sublime
uselessness." Although his interests are analytical and descriptive rather
than practical, Goffman exposes the human behavior that allows speakers
to perform well in the order of experience called the oral report or public
presentation. By showing how practiced speakers deviate from the written
speech to achieve certain effects, by demonstrating how kineseology—body
movement and gesture—lends an air of spontaneity to the event, by point-
ing out how tone, cadence, and other vocal cues *all* can be used to bind
the audience to speakers and thus facilitate the transfer of information
about the world, Goffman captures the essence of the event called *the
speech.*

Despite the fact that Goffman appears to be unmasking pretenses, seems cynically to paint speakers as slick operators capable of great deception, or worse, autonotoms locked into programmed patterns of behavior, he firmly believes "there is a structure to the world, and this structure can be perceived and reported" (Goffman, 1981, p. 195). The function of the speaker before an audience, therefore, is to present a "valid picture." Abrahams (1984, p. 83) remarked that Goffman has a "deep and ongoing reverence for the system" and that personal liberation and identity are "engineered by the ability to master the system, cracking its most important codes and practicing its most notable styles. He calls for the achieving of intellectual dominion, not for an aping of mannerly useages."

Recommendations

Goffman and Pike have argued that the lecture is a laminated affair, one that consists of different levels of composition and presentation, naturalness and artifice, reality and game. Given this predetermined mixture, speakers need to address the problems and solutions that might be employed in the two major levels: the text and the delivery.

Level 1: Preparation and Composition of the Text: Speakers as Authors

Goffman affirms that lecturers have contracted with their audiences to convey some meaningful part of the world. In other words, through organized talk, speakers can give an audience access to a picture of reality. Achieving that single goal means that even before speakers draft the first, tentative sentences, cull the first piece of evidence, or compose the first transparency, they must be convinced that they have something important to say and that they can delineate its features to the audience.

Because the author is the intellectual source of the *ideas* in the lecture, here artificiality, convention, formula, or other forms of mechanized thinking have no place. The ideas in the lecture are reflections of what Goffman (1981, p. 173) calls the "self of relatively long standing . . . the textual self." In the office or the laboratory, the executive and the scientist must think, discuss, discover, and decide through the operations of and by the rules of the textual self. The corporation or the research institution demands that the investigators ask real questions, devise effective and practical solutions, and express the issues clearly.

Thus, a speaker must confront these issues:

1. Have I determined and addressed a real question or problem, and can I state my thesis clearly and succinctly?

2. Do I actually have at my immediate disposal the facts that I can use as evidence to teach my audience what I want them to know?
3. Have I in any way distorted, inflated, or ignored factors that might alter the picture of reality I propose to project?
4. Can I elucidate a consequence that my lecture will have upon the reason, imagination, or behavior of my audience?

Stated a little less formally, speakers might ask themselves:

What do I actually know?
Do I have faith in it?
So *what* does it mean?

An honest and thorough self-examination has several benefits. First, speakers will be enthusiastic because they know that they have a product to present—the product of work, sometimes months or years of study; of insight, a new and deeper look at something; of significance, because someone in the audience will learn and perhaps act upon what they now know. Additionally, there will be at the core of speakers' lectures a certain energy that will invigorate and animate the subject. They will speak as a master talks to an apprentice in the confines of a shop.

Speakers' expertise will come through the presentation just because they know and believe in the rightness of what they have to say. When the lecture presents information and theories arrived at through the operations of the "long standing self," then the delivery will mirror the natural origin and evolution of the content. Teachers of the technical presentation know from experience that the same novice who bumbles through, for example, a simplified report on solving equations, the use of the theodolyte, or a theory of optics can strangely fascinate an audience with an account of her/his lifelong passion to breakdown a Holly carburetor, develop sepia-tone photographs, or construct a $10,000 stereo system for $99.50. All previous signs of anxiety, organizational confusion, and audience misjudgement seem to disappear. The point, once again, is that hard-nosed inquiry, intellectual preparation, and a love for the matter at hand can produce fluid communication.

Having passed the self-examination stage, lecturers must then find the form by which they present their findings. Although many current textbooks advise student speakers to construct a lecture or technical presentation just as they would a written report, Goffman explicitly rejects the notion that a speech and an essay are the same. A formal lecture should not be a verbal recitation of a written document, whether an academic study or corporate report. Instead, a lecture conforms to quite different modes or patterns of information transfer. Pace, intensity of content, level of diction, and even the overall structure of the oral report should be adjusted to meet the audience's capability to comprehend and willingness to be moved to action.

One of the most obvious differences between oral and written texts is that the listeners are asked to rely much more upon aural rather than visual powers. Yet since the Renaissance, Western people have become increasingly visually oriented, first to print and then to screen. The Rand Corporation devised the chart shown in Figure 11.1 to illustrate the effectiveness of oral and visual methods of presentation:

Method of Protection	Recall 3 Hours Later	Recall 3 Days Later
Oral only	25%	10%
Visual only	72%	20%
Blend of oral and visual	85%	65%

Figure 11.1 Briefings Communicate Effectively

In "How to Make a Better Technical Presentation," the 3M Corporation says, "When relying on verbalization alone to communicate, an estimated 90% of a message is misinterpreted or forgotten entirely. We retain only ten percent of what we hear. Adding appropriate visual aids to verbalization increases retention approximately 50%" (Visual Products, 1979, p. 5). Thus, in preparation and composition of the lecture, speakers must radically alter the proportion of words to fixed images, including pictures, diagrams, charts, graphs, symbols, and signs.

When preparing an oral report, lecturers must remember that they have an immediate relationship with the audience. Therefore, a prose introduction or conclusion, standard in a printed text, can sound ridiculous when delivered to assembled flesh and blood. Thus, the first paragraph in an essay in *The New England Journal of Medicine* does not proceed in the same manner as the opening remarks to physicians gathered for a special session at the AMA convention. Nor do speakers choose the same impersonal voice—the same deliberately distanced and formal posture—for both events. Because the physicians are present to meet and hear the figure whom Pike (1967) calls the *principal,* the authority who believes in the scripted position, they must inevitably be curious about this person. They need more time to gauge the speaker and take the measure of her/his voice, attitude, bearing, and character on the stage of the lecture hall. When technical writing texts tell speakers that the oral report has an introduction, a middle body, and conclusion, they must also describe how each major division of the speech has very different rhetorical demands and audience expectations. The parts may be familiar, but the manner of approach and follow through are surely different.

Level 2: The Delivery

Goffman's essay on the lecture should cause technical writers to steer a different course in delivering public addresses. Once we understand that the lecture is not what we would classify as a natural event, we recognize that the best public speakers are actors, people who are taking on for the moment, at least, a different voice, a different personality. In a sense, the public speaker takes on a *persona,* just as the novelist uses a created voice to address her/his readers. The narrator of a story is seldom the same as the author of a story. Experienced teachers—people who spend many hours per day speaking to an audience—know that the role they play in front of a class is a composite of their true self, the remembered image of old mentors, the idealized figure of the great speaker, and even at various times the friend, the authoritarian, and a host of other "characters."

Certainly, it seems a mistake for speakers to try to act and speak *naturally* in an *artificial* setting. Speakers feel anxious when they find it impossible to be natural under the circumstances of the public performance. In time such maturity is possible, but the command to act naturally doesn't address the immediate problem because it misses the point. Perhaps one might advise speakers to attempt to give the illusion of naturalness through divergence into the realm of *fresh talk* or through altering the cadence and the pitch or volume of the voice. These are the actor's techniques.

Much of the published material on the oral presentation has been written by teachers who have spent years concentrating on formal, standard English usage and classical patterns of rhetoric. Their experiences have made them value articulate expression, proper grammar, standard vocabulary, and precise diction. They have lived in an environment that fosters adherence to traditional rules.

Yet, their students are learning how engineers, biologists, and computer scientists talk with one another, what forms of syntax and persuasive rhetoric they consider effective, what level of diction indicates "mastery" of the field, and what pose (e.g., theatrical) to strike in public, in a seminar, or in a conference. Students and junior colleagues in science and business are continually learning more than just the content of their endeavors. Yet, many published articles on oral reports use inappropriate yardsticks to measure effective verbal communication, and this makes novice speakers anxious. They find that they must radically change their speaking styles to achieve what the textbooks define as effective communication. Again, ceremonial language must be employed in the ceremonial performance—and this language may or may not be standard or traditional in every aspect. We ought to assure these speakers that they know their audience best. Let speakers act as they learned to act.

Novice speakers should also be encouraged to feel less anxious about public speaking if they see that the lecture has a life and a form of its own and is therefore supported by a framework that cannot collapse. Although it is true that individual speakers might suffer a so-called "production crisis," a moment of anxiety and awkwardness, the whole event is still governed by the larger and stronger elements that are capable of absorbing and muffling the smaller shock waves of inexperience and nervousness. Indeed, the production crisis is part and parcel of the public speaking ritual. Could we even posit that the audience attends a lecture expecting to encounter minor lapses in speech patterns, a certain fumbling for papers, occasional fluctuations in room lighting for dramatic effect, or to hear the whirring of slide projector cooling fans? This form of acoustical and visual "entertainment," this playing with props, is common to all exercises called ceremonial. Could we similarly argue that speech flaws, idiosyncracies of pronunciation, typical haltings and stammerings make our lecturers more human, more approachable, and ultimately more effective communicators of their picture of reality?

References

Abrahams, R. D. (1984). Goffman reconsidered: Pros and players. *Raritan: A Quarterly Review,* Spring, 76–94.

Andrews, D. C., & Blickle, M.D. (1982). *Technical writing: Principles and forms* (2nd ed.). New York: Macmillan.

Geonetta, S. C. (1981). Increasing the oral communication competencies of the technological student: The professional speaking method. *Journal of Technical Writing and Communication,* **11** (3), 233–244.

Goffman, E. (1959). *The presentation of self in everyday life.* New York: Doubleday.

Goffman, E. (1974). *Frame analysis: An essay on the organization of experience.* New York: Harper & Row.

Goffman, E. (1981). *Forms of talk.* Philadelphia: University of Pennsylvania Press.

Houp, K. W., & Pearsall, T. E. (1980). *Reporting technical information* (4th ed.). New York: Macmillan.

Hymes, D. (1975). Breakthrough into performance. In D. Ben-Amos & K. Goldstein (Eds.), *Folklore: Communication and performance* (pp. 9–74). The Hague: Mouton Press.

Kimel, W. R., & Monsees, M. E. (1983). Engineering graduates: How good are they? *Engineering Education,* **74**, 210–212.

Ledger, M. (1982). The observer. *The Pennsylvania Gazette,* February, 39.

Pike, K. (1967). *Language in relation to a unified theory of the structure of human behavior*. The Hague: Mouton Press.

Schoen, R. (1979). Let's give better scientific and technical talks. *Journal of Technical Writing and Communication*, **9** (2), 163–168.

Visual Products Division (1979). *How to Make a Better Technical Presentation*. St. Paul, MN: 3 M Center.

White, J. F. (1979). Presenting Papers. *IEEE Transactions on Professional Communication*, **PC-22** (4), 179–182.

12

How Can Technical Writers Further Their Professional Development?

SANDRA WHIPPLE SPANIER

Few writers have discussed the twin concerns of professional development for technical writers and possible courses of action for practicing writers. To address this omission, Professor Sandra Whipple Spanier interviewed or corresponded with 47 professionals in the Northeast. Her correspondents were drawn from all levels of responsibility in a variety of settings from corporate to governmental agencies; they perform tasks for many different audiences and to satisfy various purposes.

Spanier found that nearly all professional, technical writers and their supervisors feel that writers' contributions to their companies have not been adequately recognized. Except in those settings where the production of written documentation is the company's product, writers were frustrated by lower salaries and status.

On the other hand, technical communicators welcome the growing sense of professionalism fostered by increased educational opportunities and a clearer perception of what qualifications technical writers need. This new professionalism is attracting to businesses writers with better, more advanced educations in technical writing. Furthermore, although management and personal directors disagree about whether a technical writer's background should be primarily in technical or humanistic areas, most technical communicators agree that successful writers are those who combine an understanding of technical subjects with versatility, good judgment, and strong interpersonal skills.

Several of Spanier's subjects recommended that technical writers combine informal, on-the-job experience with participation in seminars, conferences, or institutes. Others indicated that extensive experience in all phases of writing was the key to success. Nearly all of the respondents agreed that diversity was more desirable than specialization: writers need to understand document production from end to end. Finally, many of the respondents advised writers to project a professional image by being alert to corporate politics, taking the initiative in assisting their colleagues, defining their professional goals, exploring the possibilities for their future development as writers.

Who Are Technical Writers?

Kevin Flaherty was in Ireland, just finishing work on his Master's degree in Anglo-Irish literature, when his draft notice arrived three days before final exams. "The Army had no idea what to do with me," he says. He was put to work in public information, his first assignment to write press statements concerning pretrial hearings for the enlisted men implicated in the My Lai massacre. Later he was sent to Saigon, where he became associate editor of the newspaper distributed throughout the country and wrote speeches and scripts for the commander of the U.S. Forces. He still planned to pursue a Ph.D. in literature after the Army, but his application to graduate school "was the only piece of mail that left Viet Nam by boat," and it missed the deadline. Unable to wait a year for an income after returning home, he took a job with the Navy editing computer documents for nontechnical users. He is now an internal communications editor for a large electronics corporation, responsible for a publication distributed eight times a year to 10,000 employees and retirees worldwide—a job that requires him to work closely with engineers in order to write about highly technical subjects in nonspecialized terms.

After four years as an aircraft mechanic in the Navy, Jack Miller went back to school in the early 1950s for a private aircraft power plant license and an Associate's degree. When a leading aircraft manufacturer interviewed him after graduation, he thought it was for a maintenance job. Instead, on the basis of his technical skill, he was hired along with every other student in the top ten of his class to be a technical writer. "I'd never heard of 'technical writer,'" he says, "and I figured it would never last." But after months of "dog work"—searching for drawings, checking indexes—("really a hands-on training program"), he was given a manual of his own to prepare, and he began to enjoy the work. Two years later he took a job as a technical writer with another large aerospace firm and over the past 25 years has worked his way up to section head of the Publications Department, in charge of four group heads and 212 technical writers and editors who produce manuals for military purchasers.

About the only thing that Kevin Flaherty and Jack Miller have in common is that they both have built successful careers in technical writing. The differences in their background represent what one experienced writer calls the "classic split" in the profession—between those with liberal arts and humanities backgrounds and those who bring technical expertise to the job. That they are responsible for producing very different types of documents for very different audiences and purposes indicates the broad range of work a technical writer might do. Finally, the very fact that each of these men seems to have followed an atypical route to his present position makes him typical. "Strictly through the back door," "I stumbled into it," and "by accident" are common responses to the question, "How did you get started in your career in technical communication?" Jack Miller has never met anyone who decided "I want to be a technical writer" and then set out to become one.

To be sure, the situation is changing. Across the country the number of programs offering specialized training, certification, and undergraduate and graduate degrees in technical communication is increasing yearly. But the vast majority of those now working as technical writers and editors have had to learn on their own what skills and attributes are needed to be successful in the field and have had to chart their own courses of professional development with little help.

Furthering the professional development of technical writers is, of course, the implicit aim of every publication, conference and seminar in technical communication and the very existence of such organizations as the Society for Technical Communication (STC). But there has been little study of the subject of professional development of technical writers per se. This chapter attempts to address that issue directly and to suggest some specific courses of action that someone already working in technical communication might take in order to develop in his or her career.

These recommendations are derived from interviews with 28 technical communicators—writers, editors, and managers of technical communications groups—and questionnaire responses from 27 others with whom I was unable to meet. Most of them work in the Northeast, many on Long Island and in the New York metropolitan area. Although they by no means constitute a national random sample of the profession, they do represent a broad cross section of technical communicators in their backgrounds, levels of responsibility, work situations, and the kinds of writing and editing they do. They work in large, long-established manufacturing firms, in commercial and government-sponsored research laboratories, in fast-growing computer companies, for companies that contract with industry or government agencies to provide technical communication services, and as independent consultants. They produce manuals for military purchasers and for home computer owners, technical reports, government proposals, press releases, annual reports, and articles for employee newsletters. They have from 5 to 37 years of experience in technical communication (fairly evenly spread along that continuum) with a mean of 17.7 years. Of the 47 surveyed, 19 hold supervisory positions. Several of them have high school diplomas plus technical training or experience, usually military; four hold Ph.D.'s. The great majority have Bachelor's degrees in subjects ranging from elementary education to electrical engineering. Three have Master's degrees or are now doing graduate work specifically in technical writing. Several hold offices in professional technical communication societies, and eight teach courses in technical writing, either in-house or through a college or university.

This chapter, based on these interviews and questionnaire responses and on a review of the small body of literature that directly addresses the issue of professional development, will attempt to present a brief profile of the profession, to come to some general conclusions about the skills and attributes needed for success, and to offer some recommendations as to how one might further his or her professional development as a technical communicator.

Relevant Research: Profile of the Profession

The initial obstacle to any discussion of technical writing is the difficulty of defining the term itself. In academe, *technical writing* usually means practical writing taught to upperclass engineering and applied sciences students in order to prepare them for the work world, where they will spend a significant percentage of their time writing (Anderson, 1981; Barnum & Fischer, 1984; Davis, 1977; Spretnak, 1982). In industry, the term often means writing operations and maintenance manuals; many professional writers who deal daily with technical subjects in public relations, proposal services, or internal communication departments would not call themselves "technical writers." As one teacher of technical writing with 20 years of experience in industry put it, "You could get 20 different answers to the question, 'What is a technical writer?' "

Many have attempted to distinguish technical writing from other types of writing—by its subject matter, format, and style (Kelley & Masse, 1977; Mills and Walter, 1978), by its linguistic characteristics (Dandridge, 1973; Hays, 1961), by the type of thought process involved in its composition (Harris, 1978; Kirkman, 1963; Stratton, 1979), by its purpose or aim (Britton, 1965; Harris, 1979), and by the relationship between writer and audience (Clarke & Root, 1972; Dobrin, 1983). Even within one of these categories there are tremendous differences in approach. For example, Clarke and Root (1972, p. 18) declare that "technical writing is writing about engineering and science for a *technically trained* audience"; "science writing" is "writing about engineering and science for a *general* audience." On the other hand, Dobrin (1983) advances a highly theoretical "monadist" view that defines technical writing in terms of the complex relationships between writer, "reality," and the user.

In this chapter, I will use the term "technical writer" to mean a writer or writer/editor who presents technological material to audiences as varied as the Department of Defense, corporate executives, or computer hobby-ists—fully aware of all of the imprecisions and limitations inherent in a definition by subject matter (Harris, 1979; Walter, 1977) and the perils of either trying to draw a line between writing and editing or lumping them together.

Nearly all of those I interviewed would agree on two generalizations: that too often the importance of the technical writer's work has not been adequately recognized and that a new professionalism is emerging within the field—a change they applaud. Several expressed frustration that the technical designers get all the credit even though their work, no matter how brilliant, is useless until it is communicated with others. A former STC chapter president says, "Technical writers often have very poor per-ceptions of themselves, and unfortunately this is fostered by management." She cites an experience of a writer she knows who was employed by the federal government to prepare a manual for a system that cost $60,000 an hour to operate. Her boss would not authorize her to have the manual stitched so it would open flat rather than having it stapled together be-cause stitching would have cost an extra $200. When, in another context, I asked several internal communications writers (all with humanities back-grounds) working in an electronics corporation whether they think of themselves first as writers or as technical writers, they unanimously said "writers." Yet one added that although he calls himself a writer outside the company, he would tell anyone he met inside the company that he works in "public relations" because of the image of writing as merely a service job.

The situation seems to be different in settings where the primary product *is* information—in computer software divisions, in companies that con-tract to provide communication services, and among consultants who find that producing documents for clients or teaching writing skills to corpo-rate executives is lucrative business. These technical writers may find their communication skills more highly valued and rewarded—not only through salary, but by job title and scope of responsibility—than they might be elsewhere.

But despite any frustrations, nearly everyone agrees that the field is changing rapidly and that we are seeing an increasing professionalism in technical communications. For one thing, technical writers are more highly educated than they were 20 or even 10 years ago. In 1965, a survey of a random sample of members of the Society of Technical Writers and Publishers (a forerunner organization of STC) found that 57% held a Bachelor's degree or higher (McKee, 1969). In a 1974 STC membership survey, the comparable figure was 67%, and by 1981 it was 90% (Carter & Stolgitis, 1983). Over one-third of those responding in 1981 held a Master's degree or Doctorate; in 1974 it was about one-fourth, and in 1965, only 11.9%.

In addition, more technical writers are specifically choosing the profession and educating themselves for it. In 1971–72, The Curriculum Subcommittee of the STC's Education and Development Committee reported nine active undergraduate degree programs (Pearsall, 1973). By 1981, the STC's *Academic Programs in Technical Communication* listed 22 (Pearsall, Sullivan, & McDowell, 1981). That figure is already outdated, and the Council for Programs in Technical and Scientific Communication, whose members represent 53 institutions, is compiling a new listing.

Employers seem to appreciate the change. The director of one of the newest undergraduate degree programs has found that companies are "desperate" for well-trained technical writers and that technical communication graduates, with their professional portfolios, "knock the socks off interviewers." In fact, two juniors in his program recently were hired away by a computer company for $20,000 starting salaries on the condition that they complete their degrees—at company expense. He likens the situation to the college football draft and adds that "something is going to hit the fan quickly" when 22-year-olds with technical writing degrees start commanding salaries equal to or higher than 40-year-old company veterans.

The increasing sense of professionalism among technical communicators is also reflected in the STC's current efforts to establish a certification program. Although a 1975 inquiry of members showed little interest, a 1981 survey showed that 73% of those responding approved of the idea in principle, and an ad hoc Committee on Certification was established in 1982 (Malcolm, 1984). The committee found that 57% of employers surveyed favored a job candidate with certification over other candidates and 82% rewarded a current employee for achievement in earning certification. One member concludes that certification is "a concrete step toward professionalism" and cites the potential benefits as improved morale, improved chapter education programs and external curricula, additional society publications and manuals, motivation for self-improvement, and, possibly, compensation of the profession by industry (Manni, 1984).

A final indication of increasing professionalism is that the roles of technical writers are changing. Sanders (1982) identifies three stages of the professional status of technical writing and editing in industry. In his scheme, a company in the first stage maintains an editorial staff that mainly proofreads and does some structural editing of documents written by technical personnel. In the second stage, some employees work as writers, in contact with technical personnel throughout an on-going project, even though they may retain the title of "editor" and remain primarily responsible for editing duties. In the third stage, writers and editors are differentiated by job title and department. While the editors remain at Stage 1, the technical writers "are responsible for proposals, reports, and documentation, they work immediately with technical personnel on current projects, and they are recognized as professionals who contribute a valuable skill to the company" (Sanders, 1982, pp. 24–25).

Simply a glance through the program of the 1984 STC convention, the 31st International Technical Communication Conference, indicates the growing interest in possibilities for redefining the roles of technical communicators. Session titles included "The Technical Communicator as Professional," "Professional Development for the Writer/Editor," "Technical Editors: Are Our Roles Limited?," "Writing Documentation Before Producing the Product," and "The Role of the Technical Writer."

Qualities Needed for Success in Technical Communication

On the one hand Jack Miller, the section head of Technical Publications, says, "I can take a person with solid technical training and turn him into a writer easily." Citing two young men he has just hired out of technical school, he claims, "I'll put them under a good senior man on the floor who will teach them writing in six months." On the other hand Kevin Flaherty's boss, director of Business and Employee Communications at another large company, looks for writing skills "first and foremost" when hiring: "An intelligent person can pick up the technical material on the job in the same way he or she can learn biology or chemistry from a book in college." Because experts are readily available within the company for consultation, he feels there is actually an advantage to looking at a subject fresh and seeing it with a novice's eye when the task is to communicate technical material to nonspecialists.

The debate as to whether the word "technical" or "writer" ought to be underlined has been going on for decades within the profession (Desmond, 1984; Ennis, 1966; Hawes & Wetmore, 1983; Pearsall, 1973; Thomas & Warren, 1984; Walter, 1966; Wetmore, 1984). The answer—like so many other aspects of effective communication—seems to hinge upon the audience and purpose for one's writing (Root, 1968). But no matter which orientation they may favor in a beginner, most technical communicators would agree that in order to develop professionally in the field, a writer ultimately must have an aptitude for understanding scientific and technical subjects *and* the ability to communicate well.

Publications directly addressed to career development in technological communication offer similar outlines of the qualities needed for success, and the responses of the technical communicators I surveyed corroborate what has been said in print (Clarke & Root, 1972; Cummings, 1984; Gould & Losano, 1984). When asked what skills are needed for success in his area of technical communication, the director of a Presentation Services department, which produces reports, proposals, and visuals for an aerospace firm, replied:

> Aside from fundamental skills in handling language, in spoken and written forms, and a knowledge of associated graphics, I would say the ability to comprehend precisely *what* has to be communicated, the good judgment to know *how* a particular piece of information should be communicated, and the smarts to get the job done quickly and effectively are prerequisites for success. Lots of common sense and the ability to deal with people are also important traits. Top it off with high motivation and you've got a winner!

Although their responses varied in detail and expression, nearly everyone I surveyed touched upon four attributes needed for success in technical writing:

- Excellent communication skills.
- Technical knowledge and/or aptitude.
- Sufficient versatility and efficiency to meet the demands of the workplace (which more than one likened to the atmosphere surrounding production of a daily newspaper).
- Interpersonal and "political" skills.

And, despite the differences in their own backgrounds and in the work they do as technical communicators, they offered surprisingly consistent recommendations as to how one might further his or her professional development in the field.

Recommendations for Furthering Professional Development

Continue Your Education

Even if they had not specifically recommended it as an important factor in professional development, one could infer the importance of continuous learning simply by looking at the personal histories of technical writers in advanced positions. Carol Lof, publisher of the *IEEE Communications Magazine,* assistant to the president of the IEEE Communications Society, and immediate past president of the New York chapter of the STC, began her career in technical communications five years ago with a B.A. in Spanish, work experience as a travel agent, and no technical background. Since then she has taken 15 courses in such studies as optical fiber communications, robotics, and very large-scale integrated circuits. Jane Mitchell, head of the Technical Publication Department at AT&T Bell Laboratories, started her career 20 years ago as an engineer and consultant, with Bachelor's degrees both in mathematics and mechanical engineering. While raising a family, she worked as a free-lance engineer and as a manuscript reader for engineering societies and then entered technical communications as a writer/editor, working her way through the ranks to her present position. She has since earned an M.S. in technical communication and has taken "many seminars" in communications.

What you should continue learning about depends on the background you have already (Filowat, 1982; Mancuso, 1983; Meyers, 1983). The director of one undergraduate degree program in technical writing advises those with liberal arts and humanities backgrounds to take courses in the sciences, including one or two computer courses, and to get advanced training specifically in technical communication. Those who wish to develop professionally need to be exposed to more than just the one type or aspect of technical writing or editing they will encounter in any single job, he says. (His is one of the institutions offering a certification program for practicing technical communicators that includes training in graphics and production as well as in editing and report writing.) He advises those with strong engineering or technical backgrounds to take college or university technical writing courses plus seminars in communication skills offered through professional societies. "It's better to get blue penciling when it doesn't really count than from the boss," he says. Those among the "new breed" who have been expressly educated in technical communication "are in good shape and shouldn't have to go back to school ever" but can continue learning more informally, on the job and through reading.

Even if you have acquired the balance between technical and communication skills appropriate for your particular job, you can broaden your perspective of the profession by continuing your education. An information developer at IBM, who began his career in technical writing 20 years ago with a B.S. in electrical engineering and who now is finishing a Ph.D. in communication and rhetoric, credits attending a one-week technical writing institute for giving him his first real "feel" for what professionalism is in his field. Although he concedes that he could have gotten the *information* presented from reading journals, the personal contacts he made with other technical communicators and the exposure to broader viewpoints were invaluable. He found himself at that institute wondering how others could have a "burning feeling about their work" that he did not have then, and he was impressed by their attitude that what they were doing was a "class act." Attending an institute or seminar helps you see where the boundaries of the profession are and where you can stretch them, he says.

Continuing your education may also be necessary for career advancement or even survival. At one large company, completing a particular sequence of in-house courses is actually requisite for promotion for technical editors without previous technical training. Taking additional courses can also unofficially aid advancement. A section head of editing and production points out that "a person's willingness to give up his or her own time after work tells a lot about his commitment, and a manager may be more willing to invest in that person." The vice president of operations of a firm that contracts to provide technical communication services, himself a 28-year veteran in technical writing, notes that, for technical writers in companies like his, continuous learning is critical. His people usually are not rewriting for a target audience material provided by technical experts but are generating it themselves "from scratch," from schematics or digital diagrams, and they cannot easily consult the engineer or expert when they hit a problem. In his field of electronics, one's knowledge may be "hopelessly obsolete" even in one year, and one must keep up with state-of-the-art developments not just to advance but to maintain one's position.

Finally, many believe that computer technology will play an increasingly important role in technical writing and editing and that professional development in the future will depend partly on how well one can use the new tools of communication (Houze, 1983; Howard, 1982). A number of those I surveyed strongly recommended learning about computers, and in recent years computers have been a major topic of interest within the profession. For example, "Advanced Technology Applications" was one of five major "stems" at the 1984 ITCC and the subject of 24 of the conference sessions, and the 1982 Annual Bibliography of the Association of Teachers of Technical Writing lists 37 articles on "Technical Writing and Computers" published in that year alone (Book, 1983).

The opportunities for advanced formal education in technical communication have been increasing steadily in the past two decades (Pearsall, 1973, 1975; Pearsall, Sullivan, & McDowell, 1981). The STC's *Academic Programs in Technical Communications* (Pearsall, Sullivan, & McDowell, 1981) describes the degree programs of 28 schools, including eight graduate programs, active at that time. Although they do not purport to be exhaustive, Gould and Losano (1984) provide the most recent annotated listing, covering 17 undergraduate degree programs, 4 graduate degree programs, and 61 nondegree programs in technical communication.

As an example of a graduate course of study, the M.S. program in Technical Writing at the Rensselaer Polytechnic Institute (RPI)—established in 1955 and "the oldest in the world," according to program literature—involves 30 hours of coursework beyond the Bachelor's degree. It can be completed in one year of full-time study or several years of part-time study. The required core courses are Communication Theory, Writing and Editing, Computer Applications in Communication, Visual Layout and Design, Writing for Industry, and Writing for Publication (or Rhetorical Theory, for those who intend to teach or continue graduate study). Electives may be chosen from a variety of areas, from mass media to computer science to management to philosophy, depending on individual goals. One of the newest programs, at Miami University, leads to a Master of Technical and Scientific Communication and includes a one-semester internship in business or government that may be performed with one's employer or replaced by a thesis in the case of those who already have substantial professional experience.

Institutes, seminars, and short courses offered by academic institutions and by professional organizations also provide opportunities for professional development. Again, the longest established, begun in 1953, is Rensselaer's Technical Writers' Institute. For a week each summer, members of the RPI faculty and representatives from industry present day-long sessions on such subjects as Written Communication: Theory and Practice, Structured Writing, Editing, Graphics and Support Activities, and Training and Professional Development for Writers and Editors.

Other seminars and short courses aimed at practicing technical writers include the following:

- University of Michigan College of Engineering, Engineering Summer Conference, "Written Communications for Engineers, Technical Writers and Managers."

  One-week conference in July "designed to improve the clarity and efficiency of both persuasive and explanatory scientific and technical communication"; includes work with personal computers.
- University of Washington College of Engineering, Technical and Professional Communications Series.

  Two-day courses offered throughout the year on topics including "Grant and Proposal Writing," "Writing for Professional Journals and Trade Magazines," "Technical and Professional Writing: A Design Approach," "Writing Computer Documentation," "Technical Presentations and Briefings," "Designing Online Documentation for Computer Systems," and "Style in Technical and Professional Writing: A Practical Approach."

- University of California at Los Angeles Extension, "Professional Technical Writing Workshop."

  A one-week course offered in October.
- Massachusetts Institute of Technology, "Communicating Technical Information."

  A one-week program held in August.
- Colorado State University.

  "Communication Workshops" for individuals from business, industry, government, and science offered through the Department of Technical Journalism.

A number of others—each held for one week each summer—focus more on the teaching of technical communication and are aimed at an academic audience but may be of interest to the practicing writer as well:

- Rensselaer Polytechnic Institute, "Technical Writing Institute for Teachers" (held concurrently with the Technical Writers' Institute).
- University of Michigan School of Engineering, "Teaching Technical and Professional Communication."
- University of Minnesota Department of Rhetoric, "Advanced Seminar for Teachers of Technical Communication."
- University of Washington, Scientific and Technical Communication Program, "Teaching Technical and Professional Writing."
- Southeastern Conference on English in the Two-Year College, "Institute in Technical Communication" (held at the University of Southern Mississippi at Gulf Park).
- The Washington Center, 1101 14th Street, NW, Washington, DC 20005, "Writing in Business and Government" (features "guided fieldwork" interviewing professional writers).

For many, the most practical and accessible formal mode of continuing education may be in-house courses offered during or after work hours (Applewhite, 1983; Bulloch, 1982; Garner, 1983; Mazzatenta, 1972, 1973, 1975). Offerings vary. They may include courses in the technology important to the organization (basic electronics, for example) or in accounting or management skills as well as courses in technical communication. Depending on the company, taking an in-house course might actually be valued more highly by management than taking a university-sponsored course or seminar in that the material may be more closely tailored to the needs of the particular organization. One supervisor mentioned that increased "visibility" after hours within the company, a special benefit of in-house courses, could pay off politically.

Informal learning, however, is at least as important as pursuing more formal routes of continuing education. In fact, of the 27 technical communicators who completed a questionnaire, 21 of them rated "learning on your own about the technology important to your company or clients" either "very important" or "extremely important" as a course of action to further one's professional development—a far higher rating than they gave to taking courses or getting a degree in any of the three areas asked about: a technical field, general communication, or technical communication. The five-year regimen for professional development laid out in the STC publication *How to Prepare for the Job Beyond the Job* (Shaw, 1979) suggests not only taking at least one course a year, but supplementing that formal education with one job-related book a year, reading at least two professional journals covering some aspect of the publications field, and making it a point to learn from specialists in your company. The manager of the Technical Information Division of a scientific research laboratory believes so strongly in on-going self-education that he has asked the supervisors in his office to release every employee, paraprofessionals as well as editors, to spend one hour a day reading materials in the library. Two continuing bibliographies are useful starting places for those who wish to keep up with developments in technical communication: the annual bibliography of the Association of Teachers of Technical Writing, published in each fall issue of *The Technical Writing Teacher* since 1975, and "Recent and Relevant," a quarterly bibliography published in each issue of the STC journal, *Technical Communication,* since 1976.

Finally, simply learning by doing may be good advice for those not specifically trained as technical communicators. Ironically, one respondent who is a full-time faculty member of one of the strongest academic programs in technical communications in the nation offers two sentences of advice: "Get writing experience. Stay away from colleges." A former newspaper staff reporter with a B.A. in English has developed into a technical writer simply by writing prolifically and well about technology. Her coverage over three years of the controversy surrounding the construction of a nuclear power plant led to a prestigious award, a new assignment as environmental and scientific writer, and eventually to an advisory position in county government. She since has established her own communications consulting firm specializing in technical and scientific writing. She attributes much of her own success to having built up a portfolio of technical articles as a free-lancer in addition to her writing on the job.

Another English major also moved into technical writing through circumstance when he took a job on a university-administered government grant to write curriculum materials on technological topics. He was then asked by his department to teach a writing course for engineering students, later became director of continuing education for a national engineering society, and now is self-employed as a training consultant offering programs in corporate settings on presenting technical information effectively. His advice is this: "Work (write, teach) at explaining technology to those who don't want to understand it. The test of success is (1) their change in attitude and (2) their success in learning it. Practice, practice, practice."

Diversify Your Skills

Is it better to specialize, to become known as the person who is an expert at X, or to broaden one's range of skills? With a few qualifications, nearly everyone surveyed advised breadth over specification. "Being valuable to lots of people is good job security," as one put it. Only if you have taken a job as a specialist in a particular area is it advisable to be known for the depth of your expertise rather than your versatility. As one manager put it, "If a person is hired for a particular project and must hit the ground running with no on-the-job training, the management would be inclined to go for a specialist." But for a long-range job demanding "general problem-solving," he would prefer the "multi-faceted individual." Several in supervisory positions did speak with satisfaction of a technical writer who had developed an expertise in high-energy physics or who was "an excellent wordsmith with a wonderful sense of graphic design" or of an especially efficient "expeditor." However, they all stressed that acquiring such a reputation is desirable only *assuming* that one already possesses excellent general skills.

The advice to become knowledgeable about all aspects of production—not just to be adept at "massaging words"—echoed repeatedly in the responses of those surveyed. The editor of an internal newsletter at a large electronics corporation said, "In my publication, I'm a one-man band." "Be a pest," advises a technical information director. "Go talk to the 30-year veteran pressman and learn the trade." He maintains that to be good, technical writers need to be familiar with all aspects of the entire publication/production cycle, "from the chemistry of paper to the limits of different types of graphics." When asked if she had noticed any common characteristics among those she had seen who had *not* advanced professionally, a technical documentation supervisor at a large research laboratory cited "people whose attitude is that they're specialists, and 'That's not my job.' " She expects her writers to "take charge of the whole document, end to end, including production," and she says writers must be willing to roll up their sleeves and drive to the printer to watch the document come off the press if necessary.

Beside learning about the physical details of production, one must, also, of course, develop a broad range of writing and editing skills. "Here we're constantly switching hats," says a section head of editing and production.

His people must be versatile, not only to prepare a 4,000-page government proposal on a week's notice, but to produce marketing brochures, data sheets, an employee service manual explaining legalistic corporate procedures in everyday language, and even the company phone directory.

Shaw advises varying your on-the-job experience to supplement your present skills. If your company does not routinely provide rotating assignments for newer professional employees, he advises asking if it would be possible (Shaw, 1979, p. 8):

> If you came in as a writer or editor, ask if you could have a year or so in production. Then, if you can, ask for an assignment in advertising and sales promotion, or public relations. And a period in graphic arts, or audio visual wouldn't do you any harm. The important thing is to build up diversified experience—to learn as much as you can in a coordinated, planned program. Such experience can be invaluable to you, both in terms of professional growth and personal self-awareness.

Participate in Professional Organizations

According to a 30-year veteran technical writer, it is easy, even when working in a large corporation, to develop "tunnel vision" and come to see your company and its procedures as "the whole world." She appreciates the "more altruistic vision" of professional societies and believes that membership can enhance one's own sense of professionalism. An editing section head agrees that professional societies can broaden one's vision by "bringing up questions that you might never have even thought to wonder about on your own."

Membership is also a way to establish a professional identity and gain employer recognition. Some societies offer professional certification and most are generous with achievement awards, valuable as concrete testaments to your competence. The very fact of membership is a sign to many employers of a serious commitment to one's profession.

Through society publications, seminars, and conferences, one can keep abreast of current developments in the field. Members of the STC, for example, receive a quarterly newsletter and a journal, *Technical Communication,* that include both articles of interest to members and an ongoing bibliography of other publications in the field. The STC also publishes dozens of monographs including an anthology series, bibliographies and indexes, guides, glossaries and standards, and publications devoted to education and to professional development. The bound *Proceedings* of the International Technical Communication Conference, the annual STC convention, has been called the "Bible of the profession."

But most important are the personal contacts one makes through society membership. As one technical writer put it, "You might meet your next boss by exchanging business cards." An independent consultant says that because of her membership in a regional women's network, "jobs came out of the air." William Stolgitis, executive director of the STC, calls the membership list of his organization a "living index." It is not uncommon for a technical communication department to consist of one or two people, and a lot of "reinventing the wheel" goes on across the country. "We have authorities on every subject available," he says. "Three or four phone calls and you can find somebody who has already been through your problem."

Even a national STC officer and former chapter president concedes that while the information presented in the literature and at conference sessions is helpful, the most important benefit of society membership is meeting people informally. She finds this especially helpful to women, who may find themselves left out of the "beer and ball games" circuit in their own companies through which men often build the social networks that benefit them professionally. She notes, too, a special advantage for those early in their careers: "In a professional society, it's all first names. At a convention, a junior technical writer will find himself calling his boss 'Eric,' when otherwise he probably wouldn't even step into his office in twenty years, and certainly would call him 'Mister' if he did."

Other organizations of interest to technical writers and editors include the IEEE Professional Communications Society, the American Business Communication Association, the Public Relations Society of America, the American Medical Writers Association, and the Association of Petroleum Writers. The two primary books about technical communication as a profession each devote a chapter to professional societies and provide annotated lists of organizations, including addresses (Clarke & Root, 1972; Gould & Losano, 1984).

Polish Your Interpersonal Skills

A manager of external communications says that when he is hiring, he looks first for writing skills but that a very close runner-up in importance is what he calls "organizational skills—in the interpersonal and political sense." A technical writer or editor must be a diplomat to go back to an engineer and discuss how a journal article should be rewritten, to cajole management into agreeing to a table of contents for an internal publication, to set up interviews with technical experts, and to work with production people to get a piece out on time. One must also be alert to potential sensitivities on certain issues and be able to deal with them effectively.

When asked what they look for when deciding whether to advance a technical writer, most of those surveyed who hold managerial positions mentioned interpersonal skills. A senior public relations officer of a research laboratory says that being successful in her field is not just a matter of being a good writer but being able to work with others without ruffling feathers—able to take a firm editorial stand with a resistant scientist or engineer and also able to sense which battles should not be fought ("knowing when to give in and allow a 'utilize' "). Nor, as one might believe, are interpersonal skills important only for technical writers in such areas as public relations, marketing, or presentation services. A section head of a technical publications department that produces manuals for experts and operators, stresses that a technical writer cannot just sit at a desk and do the job: "A good technical writer has to be able to extract information from somebody. Give and take, rapport are very important." A principal software writer advises: "Learning how to get along with all sorts of people (especially the computer 'weirdos') is very useful."

When asked to what they attribute their own success in their careers, most of those I surveyed cited their ability to make a good impression and work well with others. After having had the experience of interviewing a number of "lumps" over the years, one manager who started in technical writing says he now realizes what an important factor his ability to "sell himself" had been in his own career. The deputy director of another technical communications group in the company (one of two in his entire division without a college degree—the other is the director) attributes his first promotion, from "engineering writer" to a member of a proposal group, to his "appearance"—his boss's image of him as one who could deal effectively with vice presidents, heads of departments, and "elite engineers."

Conversely, when asked to name any common characteristics they had noticed among those unsuccessful in their careers in technical communication, managers again consistently mentioned personality traits. One said he had seen "adversarial types" destroy their careers at his company. A communications manager who has risen meteorically in her organization concedes, "Others may work harder than I do, but they grumble and moan." Another is only half-facetious when he says that the key to success is "learning to work and play well with others."

Be Wise to Politics

What does it take to advance professionally in technical writing? According to the head of the Technical Publications department of a prominent research laboratory, "Dedication to excellence, a sense of humor throughout it all, and some political savvy!" Although several wrote in on the questionnaire that they hated to admit it, 18 of the 27 respondents said that "learning company organization and politics" is "very important" or "extremely important."

The very first step of the four-stage plan for professional development that Shaw outlines in the STC's *How to Prepare for the Job Beyond the Job* (1979) is to "orient yourself" to your organization within the first few months on the job. He recommends actually drawing a chart showing all the people in your office, including secretaries and clerks, along with their responsibilities and reporting relationships; circling the group leaders, supervisors, and managers in colored ink; then memorizing the details. By immediately clarifying the context in which you work, you will gain an understanding of the relationships affecting your job that many people only develop after years of experience.

"Be visible" is another bit of recurring advice. It is important that you make yourself a "known quantity," in the words of one editing supervisor. He notes that attending in-house courses and seminars and participating in company social events will demonstrate your interest in and commitment to your organization and to your work, as long as "playing politics" never overshadows your actual performance. It is also important to keep superiors aware of your continuing education efforts outside the workplace and of your achievements on the job.

"Dress for success," advises a public relations officer. "It sounds very superficial, and I would like to think it's a bunch of bull, but it isn't." In her organization, a research laboratory, "the scientists are 'It,' and if you don't have a Ph.D. in science, to some scientists you don't count, you're not worth talking to," she explains. When a writer is interviewing a scientist, it is critical to come across as a professional in order to build the mutual respect needed to work cooperatively on a piece of writing. Unfortunately, projecting a professional image is especially important for women, who are not always immediately taken seriously, she claims with regret. Others also stress the importance of appearance (Knauft, 1984; Thorne, 1982). The first of Evelyn Knauft's "Ten Commandments of Professionalism" is "Thou shalt look like a professional." She even extends her concern to the desktop: her "second commandment" is "Thou shalt maintain an office with a professional appearance."

Finally, crass as it sounds, nearly everyone agrees—but again, hates to admit—that it is critical to cultivate a good relationship with those higher up in the organization. Twenty-two of the twenty-seven who completed the questionnaire rated "developing rapport with superiors" as "very important" or "extremely important"—higher than they rated any other category except "broadening your range of skills."

"You have to have rapport with the boss," says a section head of editing and production. "If you can't get past him, you'll never get out of that area." Shaw's (1979, p. 6) fourth and final tip for professional development in technical communication is "Stay in touch with the boss." He advises talking to your supervisor frequently, not just at salary review times, and arranging planned performance reviews at set intervals if your company does not have scheduled reviews. "Make sure that you fly on the same beam with him, and that you really understand what he wants from your job."

It is important not only to get along with the boss, but to understand his or her values, priorities, and personality. "People look for reflections of themselves," a young writer cynically but probably accurately observed. Certainly it seems to be true when one compares the backgrounds of supervisors surveyed in this study with their advice to others. Many of those with graduate degrees tend to favor formal education as a means of professional development. For example, a manager who had once aspired to be either a concert pianist or physicist and who now holds two Master's degrees has little regard for in-house courses unless they are piped into an organization by closed circuit television from a university. Yet another manager, who acquired a technical background in the Navy as a signal man and quartermaster and who does not have a college degree says he would not consider university graduate courses at all when deciding whether to advance a technical editor unless that person had also completed the company's own courses in blueprint reading and electronics.

Take Initiative

"Of course, you must do your job well, but if you're *just* doing your job, you won't get the good rewards," advises an editorial manager. "Some people are willing to be robots," says another manager. "A technical writer must be a good scrounge"—he must go a step beyond what he is asked to do and must be tenacious about getting information through various routes, he explains. Most of the managers I interviewed cited initiative as an important factor in professional development. They appreciate some independence and value the technical writer "who takes the bone and shakes it," "who doesn't have to have his hand held," " who doesn't run for help until she's tried to solve the problem on her own," and "who is occasionally willing to take the risk of failing."

A consultant who was writing an Environmental Impact Statement for a land developer made her client very happy, she says, when she was able to suggest a "mitigating measure" to address the problem of nitrate levels in the groundwater. She also increased her own business since the measure she suggested later involved her preparing a brochure for future homeowners in the proposed development on suitable plantings and the use of fertilizers. She cautions that, when she was working in a corporate setting, she learned the hard way several times that one must be sensitive to politics when suggesting new procedures so that one doesn't "make an end run around the supervisor," but she generally believes that taking initiative pays off.

Management may be far more receptive to one's taking initiative than a writer or editor may realize. Several in one department rated "simply doing your job and following procedures carefully" as "extremely important," while rating "initiating new projects" as "not important." One even responded "I would advise against this." Interestingly, when interviewed, their manager praised the job they were doing but said that he wishes his editors would be "more proactive than simply reactive." On the questionnaire when asked to rate the importance of possible courses of action for furthering professional development, one respondent, a software writer who has advanced very rapidly in her five-year career, put an asterisk beside the item "simply doing your job well and following procedures carefully." She wrote at the bottom, "Doing the job well is *very* important, but that doesn't necessarily mean following procedures carefully."

Perhaps another up-and-coming young technical writer best states the case for taking initiative when he advises, "Help people do their jobs easier and faster. Nothing is as impressive as a new tool. Invent tools—smart ones—people will love you for them and pay you better, too."

Explore Possibilities for Expanding Your Role

A computer software writer who began his career in technical communication 20 years ago with a B.S. in electrical engineering speaks with excitement of the current opportunities for "REAL GROWTH in capital letters" within the profession. As Sanders (1982) notes, too, the roles of technical writers are evolving, and, at the most advanced stage of professional development in the field, writers work immediately with technical personnel at every stage of a project, helping to shape the document from the beginning. Of the 27 who responded to the questionnaire, 17 rated "becoming involved with technical people in project development before the writing-up stage" as "very important" or "extremely important." Ten of those gave it five points, meaning "extremely important." One wrote next to the item, "This one should be rated a 10!"

There are few charted paths for doing this, but, as noted earlier, the topic has generated much discussion within the profession. James A. Mann (1983), a technical writer/editor at Stone and Webster Engineering Corporation, proposes a model for expanding the writer's role that would be workable even in very traditional organizations. In this arrangement the engineer supplies information to the writer who actually writes the first draft of the paper. They meet to review it, the writer prepares a second or "final" draft, and the paper is returned to the engineer for final revision. This team approach not only results in high-quality papers while reducing the time the engineer must spend writing, it also is more challenging and rewarding to the writer than is simply cleaning up the language of a paper already written.

Others suggest pushing the writer's role even further to involve him or her directly in project development (Costello & Fenwick, 1984; Knapp, 1984). Vicky Costello, an Associate Information Developer at IBM, had programming training as well as two years of writing experience when she was assigned to write a reference manual for programmers for the command mode of a system then under development. She discovered that very little information was available from the developers, who were then concentrating on a different aspect of the project. Finally, in frustration, she asked for her own system in order to test it herself. After working with it, she was able to provide feedback to system developers and eventually knew more about the command mode than some of them did. She began to be included in the informal network of phone calls, technical discussions, and daily interactions that underlie and drive the formal communications activities of any project. The developers even requested advance copies of the manual for their own use. She and her manager conclude that besides resulting in an exceptionally high-quality manual, the arrangement increased productivity: there were no "peaks and valleys" in the writer's workload, and the review and production phase went smoothly. Because "the writer successfully penetrated the protective wall surrounding development," the vastly improved communications between writers and developers—a relationship that too often is "near adversary" (Costello & Fenwick, 1984, pp. MPD 45–46).

Nor is involvement with developers limited to those writers with technical backgrounds. Based on her experience at IBM as a member of a team redesigning an on-line database, Joan Knapp, a former technical writing teacher, acts as a liaison between developers and users in the product design stage and as a liaison among developers themselves, "helping to channel their individual efforts toward the creation of a unified system" (Knapp, 1984, p. WE-30). Dr. Knapp, project manager, was enthusiastic enough about having a writer on the design team to extoll its advantages in a company newsletter just two months after she was hired: clearer communication between designers and users, standardization among the large numbers of documents involved in the project, and increased productivity. Knapp herself extolls the benefits to the writer (1984, p. WE-33):

> I have great hope for this new avenue of technical communication. I think it has the potential not only to open up new vistas for writers but also to change the way management thinks of writers and writers think of themselves. We

know we have a great deal of potential that is often untapped in our current roles. With a set of new roles that we ourselves define, we have the opportunity to release that potential and to exert our intellectual and imaginative capabilities to improve not just communication about products but products themselves.

It should be stressed, as she says, that it often will be up to the writers themselves to initiate any redefinition of their roles.

Costello and Fenwick (1984, p. MPD-44) make the point that Costello's assignment "was simply to do whatever was necessary to write a good manual," and the developers initially reacted to her request for her own installation of the system with "good-humored surprise." It was never conceived as an experiment in expanding the roles of technical writers. But as the writer's technical involvement increased, with good results, management's interest in the arrangement grew. Knapp, too, notes that a company may not yet have realized the advantages of including a writer as an active member of a design team: "In that case, you will have to tell management about those benefits. And since you are first and foremost a communicator, you are well qualified to do so" (Knapp, 1984, p. WE-30). If you manage a team of writers, she suggests loaning one of them to a project in the design stage and as needed during implementation. The writer will return with valuable new experience; your department will gain stature in the technical community when the project manager realizes the skills a writer offers; and your department's subsequent writing tasks involving that system will be much easier with someone among you who knows firsthand what the designers intended.

It should not be forgotten that other, more traditional opportunities also exist for expanding or redefining your role. Gould and Losano (1984) note that it is important that technical writers be able to teach others about their profession. They cite a young technical writing graduate who, in addition to editing reports and classifying and abstracting patents, teaches a 13-week course in technical writing for her department. They predict that she "will go far because she has jumped into a new area and has demonstrated her ability to explain, describe, and illustrate" (Gould & Losaro, 1984, p. 18). Several of those surveyed for this chapter teach technical writing part-time in colleges and universities as adjunct faculty members. Such opportunities are likely to increase as academic programs grow. For many who are well settled in their full-time jobs, teaching adds a challenging and satisfying new dimension to their work. There are also opportunities for in-house teaching, either as a company employee or as an outside consultant (Mazzatenta, 1983; Moskey, 1983; O'Hara, 1983).

Finally, free-lancing and consulting are other alternative roles for technical writers (Handler, 1982; Holtz, 1983; Malone & Holder, 1974; Moskey, 1982). Although several said they planned to continue working in one of these capacities after retirement, those I surveyed had mixed feelings about the value of free-lancing or consulting to those early in their careers who want to further their professional development. Of the 27 who responded to the questionnaire, 4 rated free-lancing and consulting as "very important," 7 as "somewhat important," 14 as "not important," and 2 said they would advise against it. Nevertheless, it may be an attractive option for experienced writers who want to vary their work or who would prefer the flexibility these arrangements allow.

Determine What Success Means for You

Whether your ultimate goal is to be senior technical writer or a vice president, it is important to take positive steps to meet it. Shaw (1979) advocates actually planning a "personal progress campaign" with a timetable, and he maps out a four-point plan of action for the first five years of a job. A publication director for Eastman Kodak argues that you must develop the capabilities needed in the job to which you aspire "enough to indicate to management that you've got the basics and potential for a lot more *before* you're going to get promoted" (Lassiter, 1982, p. C-51). Most companies do not promote someone to a higher position and then wait for the employee to grow to fill it. One must consciously plan for the next step up.

Unfortunately, perhaps, but realistically, moving up—at least in terms of salary and job title—often means moving out. This is especially true if one works in a small department where there is little room for advancement unless someone else leaves. Of those I interviewed, nearly everyone who took a major leap in salary or responsibility did so by taking a better job offered by a different company. An internal communications editor says the best advice he ever received is not to sit too long in one place—to stay in a position only until you find you are no longer learning. "The first job is not the only job," advises another. Several mentioned that it is important to consider possibilities for professional development when choosing a job in the first place. If your company offers in-house courses or tuition reimbursement for outside education, take advantage of these opportunities—*look* for them in an employer so that you can continuously build the credentials to advance.

One fairly common career path for technical writers is to move into management positions in which, ironically, they do progressively less technical writing as they advance—much as many engineers do less and less actual engineering as they become more and more "successful." Unfortunately, the reason for this may be that, in many companies whose top priority is the technology they develop, technical writing too often is considered a peripheral, "service" function. Such companies often reward the engineer who also speaks and writes well far better than they do the good writer who also understands engineering. Technical writers may find that ceilings on their salaries and ranks are lower than those of their technical counterparts who hold equivalent degrees or who are not even as highly educated as many of the writers.

The situation may be quite different, however, in companies whose *primary* product or service is communication. A vice president of a communication services firm says that although "from a title standpoint," an administrative position may *look* like a measure of success, one does not have to go into management in his company to be financially rewarded. There is a short supply and a tall demand for the highly specialized writers he employs. In fact, the "pirating away" of good people is a constant problem. The case is similar among computer documentation writers, who may rise in a company very rapidly on the basis of the quality of their work alone, regardless of years of experience. One software writer surveyed had doubled her salary in her six-year career, and her case is not unique.

A significant proportion of those I surveyed actually were quite specific about their *not* feeling that a managerial position, in itself, is an important mark of achievement. An editorial supervisor with 27 years of experience and a Master's degree in technical communication states that point strongly: " 'Beauty is in the eye of the beholder!' Personally, I have little regard for some of the people who 'appear' to be successful in the field and/or who have achieved high management/executive positions. For most practitioners, salary rewards or accolades can never compensate for the work put in. Success can only be measured by the feeling of satisfaction that a job is done as well as possible under the conditions imposed."

Given the diversity of backgrounds, responsibilities, and work situations of technical writers—indeed, given the difficulty of even defining the term "technical writer" itself—it is hardly surprising that there are no firmly established paths to success. In fact, there is no clear-cut definition of what "success" in the profession even means. The technical communicators I surveyed offered a variety of responses to the questions, "How would you define professional advancement in this field? What are the measures of 'success'?" Only a very few mentioned salary and promotions alone; most did not mention them at all. They seemed to prefer their "pay" in the forms of peer recognition and self-esteem. Most frequently they cited recognition for work well done, increased responsibility on important projects, the acquisition of new knowledge and skills, greater independence and freedom to experiment, and self-satisfaction. Success is "to attain the

level one *wishes* to attain and one can handle well," as one technical publications department head put it.

If there is one key characteristic of those I surveyed whom most would
agree have built successful careers—at least when measured by such observable standards as salary, scope of responsibility, or rate of advancement—it is their genuine enthusiasm for their work. When asked in the
questionnaire what he hoped to be doing professionally five years from
now, a writer with a national reputation as an in-house teacher of technical communication wrote, "See answer to question #2"—"What kind of
work do you do?" A senior public relations officer for a scientific research
institution writes: "I would probably be content to write science stories at
work and a novel at home, indefinitely. Science writing is never boring.
There are new developments and discoveries, simply amazing things, at
every turn, and everyone I interview has something fascinating to tell me."
The publisher of a professional engineering society magazine who is active
regionally and nationally in the STC, when asked to what she would attribute her own success, replied simply, "I love what I do, and I love who
I do it for." Perhaps the best single bit of advice to those who want to
know how they can further their professional development in technical
writing comes from a group supervisor in the technical publications department of a large commercial research laboratory: "Choose work you
enjoy and can do, in that order."

References

Anderson, P. V. (1981). *Research into the amount, importance, and kinds of writ-
    ing performed on the job by graduates of seven university departments that
    send students to technical writing courses.* Paper presented at Modern Lan-
    guage Association Convention, New York.
Applewhite, L. (1983). An in-house editorial-tutorial program for developing com-
    munication skills. *Technical Communication, 30,* 2–4.
Barnum, C., & Fischer, R. (1984). Engineering technologists as writers: Results of
    a survey. *Technical Communication, 31,* 9–11.
Book, V. (1983). 1982 ATTW annual bibliography. *The Technical Writing Teacher,*
    11, 57–78.
Britton, W. E. (1965). What is technical writing? A redefinition. *College Compo-
    sition and Communications, 16,* 113–116.
Bulloch, W. D. (1982). A view from Bell Laboratories. In *Proceedings of the 29th
    International Technical Communication Conference* (pp. E-28–29). Wash-
    ington, DC: Society for Technical Communication.
Carter, S. G., & Stolgitis, W. C. (1983). STC membership profile. *Technical Com-
    munication, 30,* 17–19.

Clarke, E., & Root, V. (1972). *Your future in technical and science writing*. New York: Richards Rosen Press, Inc.

Costello, V., & Fenwick, W. I. (1984). The changing role of the technical writer in software documentation. In *Proceedings of the 31st International Technological Communication Conference* (pp. MPD-43–46). Washington, DC: Society for Technical Communication.

Cummings, K. (1984). Performance ranking of technical editors/writers. In *Proceedings of the 31st International Technical Communication Conference* (pp. MPD-8–11). Washington, DC: Society for Technical Communication.

Dandridge, E. P., Jr. (1973). Notes toward a definition of technical writing. *Journal of Technical Writing and Communication, 3*, 265–271.

Davis, K. C. (1980). Survey of STC membership. *Technical Communication, 27*, 13–17.

Davis, R. (1977). How important is technical writing?—A survey of the opinions of successful engineers. *The Technical Writing Teacher, 4*, 83–88.

Desmond, J. (1984). Debate continues over best skills for technical writing. *Computerworld*, May 7, 23.

Dobrin, D. N. (1983). What's technical about technical writing? In P. V. Anderson, R. J. Brockman, & C. R. Miller (Eds.), *New essays in technical and scientific communication: Research, theory, practice* (pp. 227–250). Farmingdale, NY: Baywood.

Ennis, G. J. (1966). Survey of technical writers in the aerospace industry. *STWP Review, 13*, 2–5.

Filowat, J. (1982). Career development alternatives for the non-technical person in technical communication. *Proceedings of the 29th International Technical Communication Conference* (pp. C-27–29). Washington, DC: Society for Technical Communication.

Garner, W. L. (1983). Family-run or hired gun? *Technical Communication, 30*, 5–8.

Gould, J. R., & Losano, W. A. (1984). *Opportunities in technical communications*. Lincolnwood, IL: National Textbook Co.

Handler, S. (1982). A revised model of the technical communicator: One management perspective of the freelancer. In *Proceedings of the 29th International Technical Communication Conference* (pp. C-41–44). Washington, DC: Society for Technical Communication.

Harris, E. (1979). Applications of Kinneavy's *Theory of Discourse* to technical writing. *College English, 40*, 625–632.

Harris, J. S. (1978). On expanding the definition of technical writing. *Journal of Technical Writing and Communication, 8*, 133–138.

Hawes, C., & Wetmore, B. (1983). Selecting a business writing career. In *Proceedings of the 30th International Technical Communication Conference* (pp. W&E-46–48). Washington, DC: Society for Technical Communication.

Hays, R. (1961). What is technical writing? *Word Study*, April, 1–4.

Holtz, H. (1983). *How to succeed as an independent consultant*. New York: Wiley.

Houze, W. C. (1983). Today's nontechnical writers and editors in tomorrow's 'electronic mega-cottage' world of work: Will they survive? In *Proceedings of the 30th International Technical Communication Conference* (pp. W&E-49–52). Washington, DC: Society for Technical Communication.

Howard, J. (1983). Advances in computer technology: What will the impact be for the professional communicator? In *Proceedings of the 29th International Technical Communication Conference* (pp. T-28–30). Washington, DC: Society for Technical Communication.

Kelley, P. M., & Masse, R. E. (1977). A definition of technical writing. *The Technical Writing Teacher, 4*, 94–97.

Kirkman, A. J. (1963). The communication of technical thought. *The Chartered Mechanical Engineer, 10* (11), 594–599.

Knapp, J. (1984). A new role for the technical communicator: Member of a design team. In *Proceedings of the 31st International Technical Communication Conference* (pp. WE-30–33). Washington, DC: Society for Technical Communication.

Knauft, E. A. (1984). Technology isn't all it's cracked up to be—The ten commandments of professionalism. In *Proceedings of the 31st International Technical Communication Conference* (pp. WE-26–28). Washington, DC: Society for Technical Communication.

Lassiter, K. T. (1982). Professional development for technical communicators. In *Proceedings of the 29th International Technical Communication Conference* (pp. C-49–52). Washington, DC: Society for Technical Communication.

Malcolm, A. (1984). Progress toward certification of technical communicators. In *Proceedings of the 31st International Technical Communication Conference* (pp. RET-35–38). Washington, DC: Society for Technical Communication.

Malone, M. K., & Holder, F. W. (1974). *Freelancing—A new dimension in technical communication.* Washington, DC: Society for Technical Communication.

Mancuso, J. C. (1983). Caterpillars and butterflies: Or, the process of transforming yesterday's Ph.D. in literature into today's technical writer. *The Technical Writing Teacher, 11*, 52–56.

Mann, J. A. (1983). The engineer/writer team approach in preparing a technical paper. In *Proceedings of the 30th International Technical Communication Conference* (pp. W&E-7–9). Washington, DC: Society for Technical Communication.

Manni, M. E. (1984). Historical review of certification. In *Proceedings of the 31st International Technical Communication Conference* (pp. RET-39–41). Washington, DC: Society for Technical Communication.

Mazzatenta, E. (1972). GM research course stresses involvement at all levels. In *Proceedings of the 19th International Technical Communication Conference* (pp. 59–61). Washington, DC: Society for Technical Communication.

Mazzatenta, E. (1973). In-house writing courses: Aids or anachronisms? In *Proceedings of the 20th International Technical Communication Conference* (pp. 185–187). Washington, DC: Society for Technical Communication.

Mazzatenta, E. (1975). GM research improves chemistry between science writers, editors. In *Proceedings of the 22nd International Technical Communications Conference* (pp. 153–157). Washington, DC: Society for Technical Communication.

Mazzatenta, E. (1983). A profile of in-house teachers of technical communication. In *Proceedings of the 30th International Technical Communication Conference* (pp. RET-3–6). Washington, DC: Society for Technical Communication.

McKee, B. K. (1969). Education of technical writers. *Technical Communication,* **16,** 17–20.

Meyers, R. (1983). How to convert humanitarians into technical communicators. In *Proceedings of the 30th International Technical Communication Conference* (pp. MPD-32–34). Washington, DC: Society for Technical Communication.

Mills, G. H., & Walter, J. A. (1978). *Technical writing* (4th ed.). New York: Holt, Rinehart & Winston.

Moskey, S. T. (1982). Moonlighting in the "real world": Professors as writing consultants. In *Proceedings of the 29th International Technical Communication Conference* (pp. E-77–79). Washington, DC: Society for Technical Communication.

Moskey, S. T. (1983). College instructors as writing consultants. *Technical Communication,* **30,** 9–10.

O'Hara, F. M., Jr. (1983). Hiring a private consultant. *Technical Communication,* **30,** 7–8.

Pearsall, T. E. (Ed.) (1973). University programs in technical communication. *Technical Communication,* **20,** 2–6.

Pearsall, T. E. (1975). Current university programs in technical communication. *Technical Communication,* **22,** 16–18.

Pearsall, T. E., Sullivan, F. J., & McDowell, E. E. (1981). *Academic programs in technical communication.* Washington, DC: Society for Technical Communication.

Root, V. M. (1968). Technical publications job patterns and knowledge requirements. *Technical Communications,* **15,** 5–12.

Sanders, S. P. (1982). *Technical writing in academe and in industry: A study undertaken preliminary to the proposal of a Bachelor of Science degree program in technical communications to be offered by the Humanities Department, New Mexico Institute of Mining and Technology.* Report prepared for Humanities Department, New Mexico Institute for Mining and Technology, Socorro, NM.

Shaw, J. G. (1979). *How to prepare for the job beyond the job.* Washington, DC: Society for Technical Communication.

Spretnak, C. M. (1982). A survey of the frequency and importance of technical communication in an engineering career. *The Technical Writing Teacher,* **9,** 133–136.

Stratton, C. S. (1979). Technical writing: What it is and what it isn't. *Journal of Technical Writing and Communication,* **9,** 9–16.

Thomas, M. & Warren, R. (1984). Writers without technical backgrounds: What they can offer to technical information development. *Proceedings of the 31st International Technical Communication Conference.* Washington, DC: Society for Technical Communication, WE-19-22.

Thorne, R. (1982). Prestige and professional demeanor (The success game). *Proceedings of the 29th International Technical Communication Conference.* Washington, DC: Society for Technical Communication, C-95-98.

Walter, J. A. (1966). Education for technical writers. *STWP Review,* **13,** 2–5.

Walter, J. A. (1977). Technical writing: Species or genus? *Technical Communication,* **24,** 6–8.

Wetmore, B. A. (1984). Learning the technical subject: 99 percent of the technical writer's job. *Proceedings of the 31st International Technical Communication Conference.* Washington, DC: Society for Technical Communication, WE-23-25.

Contributors

Deborah C. Andrews, professor of English and director of the business and technical writing program at the University of Delaware, teaches undergraduate writing students and conducts research on the relationship between computers and composing. She is also a private consultant for companies such as RCA-Government Systems Division, Battelle-Columbus Laboratories, AT&T Technologies, and Liebert. Andrews' teaching, research, and consulting experience inform her two books: *Technical Writing: Principles and Forms* (with M. D. Bickle) and *Write for Results* (with W. D. Andrews).

Lynn Dianne Beene, an associate professor of English at the University of New Mexico, is currently the director of the freshman writing program and a member of the professional writing program at New Mexico. As a consultant to various newspapers, government agencies, and private corporations, Beene has often been asked to explain why one document "reads better" than another. Such questions prompted her to investigate contemporary linguistic theory for its insights into document design, and this research informs much of her work and publications. She is the coeditor of *Argument and Analysis: Reading, Thinking, and Writing*.

Vera R. Charrow, former associate professor of the Document Design Center of American Institutes of Research (AIR), Washington, D.C., gained her expertise in document design through her research on legal and bureaucratic language. Currently a legal writing consultant for various companies and government agencies, Charrow is coauthor of *Clear and Effective Legal Writing*.

Robert de Beaugrande, professor of English and English linguistics at the University of Florida, is the author of numerous books including *An Introduction to Text Linguistics* (with Wolfgang Dressler) and *Text, Discourse, and Process*. In his publications, de Beaugrande is concerned with how allied fields such as artificial intelligence, cognitive science, philosophy, and psychology can contribute innovative and practical techniques for teaching language skills and for helping writers understand language dynamics more fully.

Mary Dieli, for her doctoral dissertation at Carnegie Mellon University, investigated the problems companies such as IBM and Apple had in designing documents to accompany their software. Her solution—revision filters—is one of several techniques she employs in her position as usability manager for Microsoft Corporation. At Microsoft, Dieli tests the usability of software interfaces, stand-up training, and on-line and print documentation. Among other publications, she is the author of "Reading, Writing, and the UNIX WRITER'S WORKBENCH software" in *T.H.E.: Technological Horizons in Education*.

Elaine Eldridge, assistant professor of English and technical communications at Texas A&M University, has coauthored several technical reports including "Cancer Incidence and Asbestos in Drinking Water in Western Washington." In addition to her teaching and professional writing duties, Eldridge works as a consultant for agencies such as the Fred Hutchinson Cancer Research Center and the U.W. Pregnancy and Health Program.

Michael Gilbertson, assistant professor of English at Oregon Institute of Technology, formerly directed the technical communication program at New Mexico Institute of Mining and Technology. Gilbertson did postdoctoral study in rhetoric with Frank D'Angelo at Arizona State University. He has written and collaborated on several articles evaluating graphics and technical writing that appear in journals such as *IEEE Transactions on Professional Communication, The Journal of Technical Writing and Communication,* and *The Journal of Business and Technical Communication.* Gilbertson is now collaborating on a technical writing textbook with M. Jimmie Killingsworth and Joe Chew, a graduate of New Mexico Institute's writing program.

V. Melissa Holland, formerly an editor and writer for the Document Design Center of the American Institutes of Research (AIR), now works as a research psychologist and writer for the Army Research Institute for Behavioral and Social Sciences in Washington, D.C. Holland has written on the application of cognitive research to problems in training and education and has conducted studies with Veda R. Charrow and William W. Wright for AIR on the effective design of public documents and procedural instructions.

Michael P. Jordan, professor at Queen's University, Alberta, Canada, understands the common problems engineers and writers face because he is a professional engineer and administrator with experience as a technical writer and publications manager. Perceiving a lack of scholarship in both the theory and practice of teaching writing, Jordan has attempted to improve this situation in his numerous articles and books on technical writing, business communication, English as a second language, linguistics, and instructional science. He is widely known on both sides of the Atlantic as a language scholar and has received various awards including the 1982 N.C.T.E. award for best scientific and technical communication research article.

M. Jimmie Killingsworth, associate professor of English, directs the professional writing program at Memphis State University. He has published articles on composition, literature, and technical writing in such journals as *College English, The Journal of Technical Writing and Communication,* and *IEEE Transactions on Professional Communication.* With his coauthor Michael Gilbertson, Killingsworth is designing a project to apply rhetorical and psychological theories to the teaching and practice of technical writing and is collaborating on a technical writing textbook.

Scott P. Sanders, associate professor of English at the University of New Mexico, argues persuasively in his role as director of professional writing for the importance of academic and professional training in writing. Recognizing that the academic and professional communities can enrich each other, Sanders had worked as a consultant and editor presenting seminars in technical writing for several firms in New Mexico and Colorado. In addition, as a summer university faculty appointee, he was a staff writer for the Sandia National Laboratories' employee publication, *The Lab News*. A frequent contributor to several journals, Sanders is also the associate editor in charge of communication education and training for the *IEEE Transactions on Professional Communication*.

David H. Smith completed a baccalaureate degree in physics and combined his interests in science with writing in the graduate program in English at the University of Delaware. At Delaware, Smith used his computer to write treatises about the history of science and about using a computer to write treatises. After completing his graduate studies, Smith joined the faculty of the Central Institute of Finance and Banking in Beijing, China, where he combined his background in science, history, and third-world economics with linguistic analyses of government documents.

Sandra Whipple Spanier, assistant professor of English, received her doctorate from Pennsylvania State University and taught technical writing to engineering students, practicing engineers, and managers in the department of technology and society for the College of Engineering and Applied Sciences at the State University of New York at Stony Brook before joining the faculty at Oregon State University. She is the author of *Kay Boyle: Artist and Activist*. In addition to her literary publications, Spanier has also conducted studies of the on-the-job writing practices of engineers and applied scientists.

Peter White, associate professor of English and American studies at the University of New Mexico, studied technical writing at the Pennsylvania State University with Professor Kenneth Houp and directed the technical writing program at New Mexico for six years. In addition to his publications on early American literature and the folklore of New Mexico, White has served as a technical writer and editor for Kirkland Air Force Base and has worked as a consultant on technical and business writing for local industries and corporations.

Joseph M. Williams, professor of English and linguistics at the University of Chicago, brings to his discussion of language and style both academic experience and professional expertise as a senior partner in Clearlines, a consulting firm that has provided writing programs for several major corporations, governmental agencies, and law firms both in the United States and Europe. Williams is the author of several analyses of language, including the bestselling *Style: Ten Lessons in Clarity and Grace,* and is coeditor of *Style and Variables* and *Standards and Dialects*. He is currently completing an introduction to legal writing, a second edition of his textbook *The Origins of the English Language,* and a coauthored study of discourse structure.

William W. Wright, a project director and editor at the Document Design Center of American Institutes of Research (AIR) in Washington, D.C., designed, developed, and tested technical publications and on-line materials for companies such as IBM. Before joining AIR, Wright used the experience he gained at the Bread Loaf School of English, Middlebury College, to initiate a national microcomputer-based communications network of teachers and writing specialists. In addition to his other publications, Wright is a frequent contributor to *Simply Stated,* the Document Design Center's newsletter.

Index